United States Nuclear Regulatory Commission

Protecting People and the Environment

NUREG/CR-7160
SAND2012-3144P

I0482672

Emergency Preparedness Significance Quantification Process: Proof of Concept

Office of Nuclear Security and Incident Response

AVAILABILITY OF REFERENCE MATERIALS
IN NRC PUBLICATIONS

NRC Reference Material

As of November 1999, you may electronically access NUREG-series publications and other NRC records at NRC's Public Electronic Reading Room at http://www.nrc.gov/reading-rm.html. Publicly released records include, to name a few, NUREG-series publications; *Federal Register* notices; applicant, licensee, and vendor documents and correspondence; NRC correspondence and internal memoranda; bulletins and information notices; inspection and investigative reports; licensee event reports; and Commission papers and their attachments.

NRC publications in the NUREG series, NRC regulations, and Title 10, "Energy," in the *Code of Federal Regulations* may also be purchased from one of these two sources.
1. The Superintendent of Documents
   U.S. Government Printing Office Mail Stop SSOP
   Washington, DC 20402-0001
   Internet: bookstore.gpo.gov
   Telephone: 202-512-1800
   Fax: 202-512-2250
2. The National Technical Information Service
   Springfield, VA 22161-0002
   www.ntis.gov
   1-800-553-6847 or, locally, 703-605-6000

A single copy of each NRC draft report for comment is available free, to the extent of supply, upon written request as follows:
Address: U.S. Nuclear Regulatory Commission
         Office of Administration
         Publications Branch
         Washington, DC 20555-0001
E-mail: DISTRIBUTION.RESOURCE@NRC.GOV
Facsimile: 301-415-2289

Some publications in the NUREG series that are posted at NRC's Web site address http://www.nrc.gov/reading-rm/doc-collections/nuregs are updated periodically and may differ from the last printed version. Although references to material found on a Web site bear the date the material was accessed, the material available on the date cited may subsequently be removed from the site.

Non-NRC Reference Material

Documents available from public and special technical libraries include all open literature items, such as books, journal articles, transactions, *Federal Register* notices, Federal and State legislation, and congressional reports. Such documents as theses, dissertations, foreign reports and translations, and non-NRC conference proceedings may be purchased from their sponsoring organization.

Copies of industry codes and standards used in a substantive manner in the NRC regulatory process are maintained at—
       The NRC Technical Library
       Two White Flint North
       11545 Rockville Pike
       Rockville, MD 20852-2738

These standards are available in the library for reference use by the public. Codes and standards are usually copyrighted and may be purchased from the originating organization or, if they are American National Standards, from—
       American National Standards Institute
       11 West 42nd Street
       New York, NY 10036-8002
       www.ansi.org
       212-642-4900

NUREG/CR-7160
SAND2012-3144P

Emergency Preparedness Significance Quantification Process: Proof of Concept

Manuscript Completed: August 2012
Date Published: June 2013

Prepared by
Randolph Sullivan
Joseph Jones*
Jeff LaChance*
Fontini Walton*
Scott Weber*

*Sandia National Laboratories
 Albuquerque, NM 87185

Operated by
Sandia Corporation
For the U.S. Department of Energy

Prepared for
Division of Preparedness and Response
U.S. Nuclear Regulatory Commission
Washington, DC 20555-0001

R. Sullivan, NRC Technical Lead

NRC Job Code R3149

Office of Nuclear Security and Incident Response

Sandia National Laboratories is a multi-program laboratory managed and operated by Sandia Corporation, a wholly owned subsidiary of Lockheed Martin Corporation, for the U.S. Department of Energy's National Nuclear Security Administration under contract DE-AC04-94AL85000.

ABSTRACT

In an ongoing effort to increase effectiveness and efficiency through improved prioritization of regulatory activities, a decision process has been developed to aid in the determination of risk significance of Emergency Preparedness (EP) program elements. The DedUctive Quantification Index (DUQI) method was developed and used in a proof of concept application for two representative nuclear power plant sites. The results show the cumulative population dose is reduced through implementation of a formal EP program compared to conditions in which an emergency response would be implemented in an ad hoc manner. Dose was shown to be consistently lower for all analyses. The DUQI method was also applied to determine risk significance of specific EP elements. Analyses included a response where sirens are assumed not operable in the 2-5 mile area around the nuclear power plant, and for a delay in notification to offsite response organizations. Detailed consequence analysis modeling was performed using site specific information. The process used information from historical studies, such as NUREG-1150 combined with current knowledge. Data for specific sites was used in selected areas to increase the credibility of the product, but the results are not applicable to any specific site. Improvements were made to the modeling approach by simulating evacuee road loading in greater detail than previous studies. The 95th percentile cumulative population dose results were produced and used to support the study conclusions.

TABLE OF CONTENTS

LIST OF FIGURES

LIST OF TABLES

EXECUTIVE SUMMARY

In an ongoing effort to increase effectiveness and efficiency through improved prioritization of regulatory activities, a decision process has been developed to aid in the determination of risk significance of Emergency Preparedness (EP) program elements. The **D**ed**U**ctive **Q**uantification **I**ndex (DUQI) method was developed and used in a proof of concept application for two representative sites with two accident sequences at each site. The DUQI method potentially provides a means to risk inform regulatory oversight of nuclear power plant EP programs. The results of this project will allow the staff to determine whether or not it is appropriate to propose policy changes for the EP planning basis, regulations, and/or guidance.

Initial analyses were performed to compare the potential consequences of accident scenarios when a radiological emergency response plan is fully implemented and when a basic all

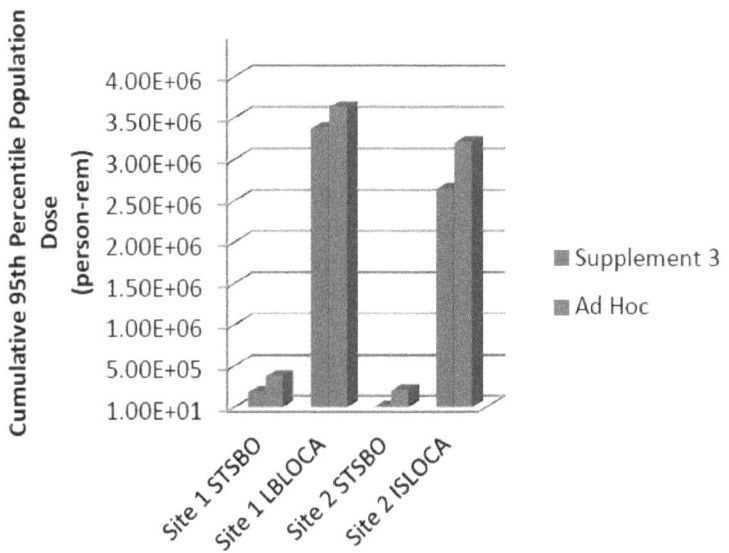

Figure ES-1. Cumulative Population Dose for the Supplement 3 and Ad Hoc Response

hazards response plan, not specific to a radiological emergency, is implemented (e.g., an ad hoc response). The results illustrated in Figure ES-1 show the cumulative population dose is reduced when implementation of a formal EP program is in place. Cumulative population dose to the public was shown to be lower for all scenarios in which an EP program was implemented. These results quantify the value of EP in terms of dose avoided by the public through implementation of an EP program. Through these results, the project has also shown one approach to risk informing regulatory oversight of EP.

The DUQI method was then used to determine whether the risk significance of specific EP elements could also be quantified. Analyses were completed for a response where sirens are assumed not operable in the 2-5 mile area around the nuclear power plant (NPP). Analyses were also completed for a response in which a delay in notification to offsite response organizations (OROs) was assumed. Data for specific sites was used in selected areas to increase the validity of the results, but results are not directly applicable to any specific site. The large number of cohorts and the approach to modeling for this project represents the highest fidelity use of the MACCS2 consequence code ever attempted. The 95th percentile cumulative population dose results were produced and used to support the study conclusions.
The results for Sites 1 and 2 are presented in Table ES-1 and show that for Site 1, a one-hour notification delay increases the cumulative population dose by about 20 percent.

Table ES-1. Site 1 and 2 Comparison of EP Elements to Baseline Results

Scenario	Site 1 (person-rem)	Site 2 (person-rem)
Baseline	1.78×10^5	1.65×10^3
Notification Delay – Entire EPZ	2.12×10^5	3.90×10^3
No Sirens (2-5 miles)	1.93×10^5	1.95×10^3

The delay in response due to no sirens in the 2-5 mile area also shows an increase in dose, but this is not as great as the notification delay. The results for Site 2 show that a one-hour notification delay increases the dose by more than a factor of 2. The delay in response due to no sirens in the 2-5 mile area also shows an increase in dose for this site. This analysis shows that risk significance of EP elements can be quantified.

It is interesting to note that a delay in notification of the EPZ public could be due to untimely classification, notification, protective action recommendation development, protective action decision making or failure of equipment. This delay is more significant than a localized failure of sirens due to the effectiveness of backup notification measures, societal notification and low population density in the cases analyzed.

The use of risk information can help prioritize resources while enhancing focus on safety, increasing public confidence, and reducing unnecessary regulatory burden. This project has shown the potential to determine the risk significance of EP program elements. The DUQI method could potentially also be used to determine the risk significance of mitigative actions.

ACRONYMS

ANS	American Nuclear Society
ASME	American Society of Mechanical Engineers
BWR	Boiling Water Reactor
CDF	Core Damage Frequency
CFR	Code of Federal Regulations
DBE	Design Basis Event
DUQI	**DedUctive Quantification Index**
EAS	Emergency Alert System
ECCS	Emergency Core Cooling System
EDMG	Extreme Damage Mitigation Guide
EOP	Emergency Operating Procedures
EP	Emergency Preparedness
EPZ	Emergency Planning Zone
ETE	Evacuation Time Estimate
ETE90	90 Percent Evacuation Time Estimate
ETE100	100 Percent Evacuation Time Estimate
FEMA	Federal Emergency Management Agency
HPCI	High Pressure Coolant Injection
ICRP	International Commission on Radiation Protection
ISLOCA	Interfacing Systems Loss of Coolant Accident
KI	Potassium Iodide
LBLOCA	Large Break Loss of Coolant Accident
LER	Large Early Release
LERF	Large Early Release Frequency
LOCA	Loss of Coolant Accident
MACCS2	MELCOR Accident Consequence Code System Version 2
NPP	Nuclear Power Plant
ORO	Off-site Response Organization
OSC	Operational Support Center
PAG	Protective Action Guides
PAR	Protective Action Recommendation
PRA	Probabilistic Risk Assessment
PRT	Pressurizer Relief Tank
PWR	Pressurized Water Reactor
RBR	Enhanced Emergency Planning Report
RCIC	Reactor Core Isolation Cooling
ROP	Reactor Oversight Process
RPV	Reactor Pressure Vessel
SAMG	Severe Accident Management Guide
SBO	Station Blackout
SGTR	Steam Generator Tube Rupture
SIP	Shelter in Place
SORV	Stuck Open Relief Valve
SRV	Safety Relief Valves
STSBO	Short Term Station Blackout
TDAFW	Turbine Driven Auxiliary Feedwater
TSC	Technical Support Center

1.0   INTRODUCTION

The science of nuclear power plant accident analysis has progressed to the point that it is now possible to enhance the emergency preparedness (EP) regulatory structure with risk-based information informed through consequence analyses. This study explored the potential to risk inform EP regulatory oversight and showed that a suite of credible scenarios important to emergency planning could be used for regulatory oversight. The techniques developed in NUREG/CR-6953, "Review of NUREG-0654, Supplement 3, 'Criteria for Protective Action Recommendations for Severe Accidents,'" Volumes 1 (NRC, 2007) (hereinafter referred to as the PAR study) and in the State of the Art Reactor Consequence Analyses (SOARCA) project (NRC, 2012a and 2012b) informed this study.

Regulatory oversight is, in part, maintained through critique of performance and review of the corrective action system. Enhancement of regulatory oversight is pursued when advancements in technologies, knowledge, etc., suggest that benefits may be achieved. In 2000, the NRC updated the EP regulatory oversight regimen to include performance measures. The update of the NRC Reactor Oversight Process (ROP) focused inspection on risk-significant areas of EP and created a "licensee response band" to allow nuclear power plant (NPP) operators to resolve issues with low regulatory significance without additional regulatory oversight. Another significant rulemaking effort to enhance EP requirements was finalized in December 2011. While the existing regulatory oversight regimen is protective of public health and safety, this study considers whether a more analytical treatment of the EP regulatory structure could be practical and beneficial. The risk significance approach diagrammed in Figure 1-1 was applied using a cumulative population dose metric to examine consequences conditional on an accident, rather than conditional on core damage, which is a more typical application.

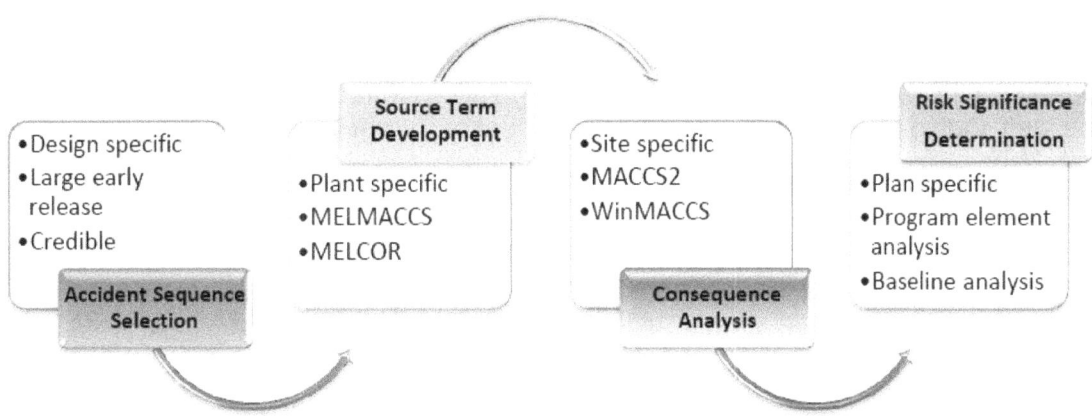

Figure 1-1. Risk Significance Determination Process to Inform EP Regulatory Oversight

The goal of this study was to determine whether a technical basis can be developed to support a new regulatory regimen that is more risk informed. The study quantifies the value of EP in terms of dose avoided. The **D**ed**U**ctive **Q**uantification **I**ndex (DUQI) method was developed to quantify the "value" of EP program elements. Understanding the value of EP at the element level will facilitate focusing resources on the most risk-significant elements. It is envisioned the DUQI method could be used in significance determination for noncompliance issues and to make quantified statements of the protection goals provided by EP, such as:

- In a severe nuclear power plant accident, there will be no early fatalities among the public who follow protective action direction;
- In a severe nuclear power plant accident, there will be no measurable increase in health effects among the public who follow protective action direction;
- In a severe nuclear plant accident, dose will be minimized among the public who follow protective action direction;
- In a severe nuclear power plant accident, no member of the public who follows protective action direction will receive a dose greater than 'X' rem (X will be a site specific value);
- In a severe nuclear plant accident, offsite contamination will be minimized through coordinated mitigation efforts.

Such statements would be based on consequence analyses that fully integrate EP elements such as: evacuation plans, event response, mitigation capability, public warning systems, protective action logic, etc. Potentially, quantitative analysis could provide an alternative to the "reasonable assurance standard" currently used for communicating adequacy of EP programs.

The NRC has pursued the goal of enhancing its regulations through the use of risk informed and performance based methods. In a June 1, 2006, Staff Requirements Memorandum (M060503B) the Commission directed the staff in part, as follows:

> "...The staff should improve the risk-informed regulation implementation plan (RIRIP) so that it is an integrated master plan for activities designed to help the agency achieve the Commission's goal of a holistic, risk-informed and performance-based regulatory structure. The plan should continue to give priority to risk-informed activities underway and incorporate lessons learned from earlier activities as appropriate.
>
> The staff should look for opportunities to enhance interactions with stakeholders as the staff moves forward with risk-informing NRC regulations and other regulatory processes. The staff should give priority to the development of such regulations and processes most likely to be utilized. The staff should ensure that processes are in place to resolve issues in a timely manner, including raising issues to senior management and/or the Commission, as appropriate.
>
> The staff should seek ways to communicate the purpose and use of PRAs in NRC's reactor regulatory program more transparently to the public and stakeholders."

The method explored in this study would support the development of a risk informed and performance based regulatory regimen for EP oversight. The product would be used to work with licensees, local communities, and the Federal Emergency Management Agency (FEMA) to begin the next major EP enhancement which would quantify the protection that EP plans and procedures should achieve and codify them in regulations that are transparent, objective, and measurable.

1.1 Scope

This study explored the potential to risk inform EP regulatory oversight through development of a risk significance determination tool, the application of which is conditional on the specified accident sequences. The project identified a spectrum of accident scenarios important to emergency planning to show how important scenarios, appropriate for regulatory oversight, may be used to inform EP oversight.

The scope included developing a method to quantify the protection EP provides and a system to determine the risk significance of EP program elements. To accomplish this scope, it was necessary to establish a baseline analysis for comparison. Two primary response conditions were established including:

- Supplement 3 Response – The Supplement 3 response, for the purpose of this study, implies that the initial response by the onsite and offsite emergency response organizations (EROs) will follow the structured radiological emergency response program. For onsite EROs, this includes following all regulatory requirements and guidance for emergency planning and implementing the guidance provided in Supplement 3 with regard to protective action recommendations. For offsite EROs, this includes implementing an offsite radiological emergency response program.

- Ad hoc Response – An ad hoc response, for the purpose of this study, implies there is no formal radiological emergency response program onsite or offsite. Although this concept is difficult to envision after decades of existing radiological emergency response programs, it is necessary to establish the differences that exist when a program is in place compared to if a program were not in place. For example, in the ad hoc response, the analysis assumes there are no designated emergency action level requirements to classify and notify offsite EROs in the event of an accident and that the offsite EROs would eventually be informed. However, the analysis assumes the plant would identify the accident and respond onsite to mitigate the accident. Similarly, the ad hoc response assumes that the offsite ERO has an all hazards emergency response plan and has the capability to respond, but does not have any radiological training or preplanned response activities such as sirens for prompt notification, preplanned traffic control, or prescripted EAS messaging.

The study models two accident sequences at two reference sites comparing emergency plan implementing procedures using Supplement 3 guidance and an ad hoc response. The difference in consequences between the Supplement 3 emergency response scenario and the ad hoc response scenario represents the value of the EP program, given a set of severe accident scenarios. Additional analyses included evaluation of consequences when an EP element is removed from the system and of an imposed delay in the emergency response timeline. More specific scope elements included:

- Review of existing probabilistic risk analyses for two reference sites to identify credible scenarios, including hostile action;
- Review of the technical basis for risk informing EP presented in the Enhanced Emergency Planning RBR report (RBR, 2007). Performing and comparing similar calculations;
- Performing calculations using important scenarios to illustrate the effect of response measures for two sites;
- Quantifying the protection of EP plans;
- Developing a proof of concept to quantify risk significance of EP program elements to inform regulatory oversight of EP.

1.2   Objective

The objective of this study was to determine through a proof of concept whether an analytical technique can be developed to risk inform EP oversight. The method would provide the regulatory basis to:

- Quantify the protection provided by EP programs;
- Support development of a significance determination process for individual EP program elements;
- Advance modeling of emergency response to identify risk significant elements; and
- Support development of a risk informed and performance based regulatory oversight regimen.

1.3   Background

The NRC EP regulatory oversight regimen was promulgated in Title 10 of the U.S. Code of Federal Regulations (CFR) in August 1980. Those rules were drafted, offered for public comment, revised, and issued 18 months after the Three Mile Island NPP accident. Early studies of severe reactor accidents and their consequences led NRC staff to conclude that the preferred initial protective action recommendation (PAR) for a severe accident is to evacuate promptly, rather than have the population shelter-in-place (SIP). Licensees are required to ensure the capability exists to notify the public of the urgent need to take protective actions within about 45 minutes of event classification. These requirements are based in part on the analysis of WASH-1400 that core damage could develop in as little as 30 minutes. The NUREG-1150 analyses supported this basis (NRC, 1990).

Licensees are also required to establish a 10-mile plume exposure pathway emergency planning zone (EPZ). The technical basis for the EPZ is provided in NUREG-0396 (NRC, 1978) and considers the likelihood of large radiological releases, the radionuclide inventory available for release and the timeliness of emergency response. Supplement 3 to NUREG-0654/FEMA-REP-1, Rev.1 (Supplement 3) provides guidance for implementing protective actions through a simplified decision-making process (NRC, 2011a).

In 2002, the staff analyzed adequacy of the EP planning basis given the change in the threat environment following the events of September 11, 2001. It was concluded that hostile action could not cause a reactor accident that occurs faster or is larger than that addressed in the planning basis. In July 2004, the NRC initiated a project entitled, "Review of NUREG-0654, Supplement 3, "Criteria for Protective Action Recommendations for Severe Accidents," (NRC, 2007). The objective of the project was to provide an evaluation of the current NRC PAR guidance contained in Supplement 3. The "PAR Study", as it has come to be called, resulted from Commission direction that the effectiveness of the NRC's PAR development guidance be reviewed and the relative merits of certain protective actions (e.g., evacuation and sheltering-in-place) be studied for a variety of situations. Supplement 3 was updated in 2011 (NRC, 2011a) based on the results of the PAR Study. The updated guidance makes broader use of SIP and provides guidance on protective actions for response to large early releases (LER) or "fast breaking" emergencies. It is this guidance that was used in the development of protective action parameters used in this study. Reviews and studies, such as those identified, help advance the knowledge of EP which is regulated under 10 CFR 50 "Domestic Licensing of Production and Utilization Facilities."

In SECY-06-0200, "Results of the Review of Emergency Preparedness Regulations and Guidance," (NRC, 2006) staff recommended a series of changes to the existing EP regulations and guidance based upon analysis of issues. As the EP program has matured, the staff recognized the benefits of a performance-based regulatory structure and conceptualized the basis for a voluntary performance-based EP regulatory regimen which could be adopted in lieu of the existing EP regulations contained in 10 CFR Part 50. The current regimen tends to emphasize compliance with, and control over, emergency plans and facilities. The performance-based regimen would focus licensee efforts on actual performance competencies, rather than maintenance of emergency plans and procedures. The performance-based regimen would provide the NRC with enhanced oversight of the actual competencies important to protection of public health and safety while allowing licensees increased flexibility. The performance-based regimen would also be supported by a set of performance indicators that would measure emergency response performance in the period between drill/exercise inspections (NRC, 2006).

In July 2009, an unsolicited industry report was submitted to NRC presenting what was described as a technical basis for risk informing EP by quantifying consequences of various response actions to severe accidents. The "Enhanced Emergency Planning" report by RBR Consultants, Inc., assessed selected hostile action scenarios as bounding cases for emergency response (RBR, 2007). The scenarios involved rapid releases that would be considered LERs. The report suggests that protective actions could be modified to focus on areas close to the plant and rely solely on SIP for areas further away. As a result of staff research, Commission direction, and in some measure the RBR report, this project was initiated to explore the potential to determine the risk significance of EP program elements for use in regulatory oversight.

1.4   Approach

Two reference sites were used for this demonstration effort. Reference Site 1 is a pressurized-water reactor (PWR) at a high population density site. Reference Site 2 is a boiling-water reactor (BWR) at a medium population density site. Medium and high population density sites were selected because they typically have longer evacuation time estimates (ETEs) and correspondingly slower evacuation travel speeds than low population density sites. A longer timeframe to implement an evacuation protective action was expected to provide more applicable information for this study than a low population density site might provide. Both sites are located in the eastern United States. Actual meteorological data was used for each site and onsite and offsite response activities were developed using site specific information.

Two accident scenarios were selected for each site and estimates of the magnitude and timing of releases were developed. Consequence analyses were performed to calculate the cumulative population dose under the postulated conditions. To determine whether the value of EP could be quantified, a baseline EP model was developed. The baseline model assumed successful implementation of emergency plans using the Supplement 3 PAR Logic Diagram (NRC, 2011a) and was compared to analyses for the assumed condition that an EP program was not in place. The difference in cumulative dose was calculated to establish the value of EP. The dose metric for use in determining risk significance in this project was selected based on the International Commission on Radiation Protection (ICRP) Publication 103, "The 2007 Recommendations of the International Commission on Radiological Protection," (ICRP, 2007). Section 4.4.7 "Collective Effective Dose," and Section B.5.9, "Collective Dose," of ICRP Publication 103 explain that collective dose may be used for optimization purposes for a specific range in time and space. In this analysis, only the EPZ and a seven day emergency phase period are considered for cumulative population dose.

2.0   ACCIDENT SEQUENCE SELECTION

The accident sequence selection for this project used information from past Level 3 Probabilistic Risk Assessments (PRAs) combined with more current accident frequency and consequence analyses to identify a spectrum of severe accidents that is appropriate for use in developing an NPP regulatory oversight regimen. The documentation reviewed included existing PRA documents, Individual Plant Examinations, and the Standardized Plant Accident Response (SPAR) models used for regulatory compliance issues. In addition, broader perspectives for BWRs and PWRs were obtained from NRC and industry studies that generated a list of credible accidents for use in establishing emergency response. These studies include NUREG/CR-6953 (NRC, 2007), EPRI-1015105 (EPRI, 2007), SOARCA (NRC, 2012a and 2012b) and other documents that address consequences from severe accidents in existing light-water reactors.

The study team also undertook a broad review of the types of accident sequences that are important with respect to various risk measures, including core damage, containment failure, and source terms. The majority of the information used is related to the two reference plants chosen. A complete description of the sequence selection process is provided in Attachment 1, "Draft Letter Report: Accident Sequence Selection."

2.1   Accident Sequence Selection Criteria

Criteria were established for the selection of a spectrum of accidents. Probabilistic and deterministic related criteria were identified. Deterministic criteria include the timing and magnitude of potential radionuclide releases. Only accidents that result in relatively early radiological release are important to this project from an emergency response perspective. Probabilistic criteria were used to eliminate scenarios that do not have a credible frequency of occurrence, even though they may result in significant releases. Frequency criteria were established to address the frequency of accident initiating events, accident sequences resulting in core damage, and the frequency of radioactive release. Random, internal initiating events that are very low in frequency were eliminated from consideration. Similarly, extremely unlikely external hazards were also eliminated. Typically, PRAs use an initiating event frequency and hazard truncation value of 1×10^{-7}/year (yr). The American Society of Mechanical Engineers/American Nuclear Society (ASME/ANS) PRA standard (ASME/ANS, 2009) indicates that this is an acceptable screening value. This approach to selection criteria is appropriate for this demonstration project, a more formalized criteria might be applied if the system advances to regulatory oversight.

NPPs use many safety systems designed to mitigate accident scenarios. Non-safety systems are also available for accident mitigation. Although an accident initiator may have a relatively high frequency of occurrence, mitigating systems reduce the potential for core damage and radioactive release. PRAs evaluate the potential for failure of mitigating systems following accident initiating events that could result in core damage or radioactive material release. Level 1 PRAs evaluate the potential for core damage, and Level 2 PRAs extend the analysis to the evaluation of radioactive release. Most existing PRAs are Level 1 PRAs and thus only evaluate core damage frequency (CDF) and large early release frequency (LERF) because these are two metrics used in current risk-informed regulatory applications. The NRC uses a CDF value of 1×10^{-6}/yr and an LERF value of 1×10^{-7}/yr in regulatory guidance, such as Regulatory Guide 1.174, (NRC, 1998a) as a threshold for non-significant changes with respect to CDF and LERF, respectively.

International and U.S. standards were reviewed for consideration of an accident frequency truncation value appropriate for use in risk-informing EP oversight. A 1×10^{-7}/yr criterion was recommended for all levels of accident delineation (i.e., core damage sequences to accident progression bin frequencies). This relatively low criterion is equal to or below most criteria currently in use in the U.S. and abroad and is recommended for use in eliminating accident scenario types from consideration in an EP regulatory oversight regimen.

Accident scenarios can be initiated by random failures, external hazards, and hostile action (e.g., armed attack). These events can occur while the plant is at power, shut down, or refueling. The magnitude of the radioactive release, the timing of the release, and the potential for affecting emergency response differ for each scenario. A credible spectrum of accident scenarios should encompass different plant operating states and hazards.

The set of credible accident scenarios does not necessarily bound the worst case imaginable, but represents a set of scenarios appropriate for regulatory oversight purposes. A truncation value of 1×10^{-7}/yr was used in identifying a spectrum of accidents with early release of various magnitudes. The following additional criteria were used in selecting the appropriate accident scenarios:

- Accident sequences that can be caused by random failures, external events, or hostile actions should be selected to reduce the number of scenarios requiring detailed evaluation;
- Accident sequences that provide similar source terms for both PWRs and BWRs and for different operating ranges should be considered in order to reduce the number of scenarios requiring evaluation;
- The accident sequences should reflect important scenarios for similar plant types and, to the extent possible, all PWRs and BWRs;
- The selected scenarios should reflect the most recent information available with regard to frequency and importance to risk;
- Early release sequences should be emphasized as they provide the greatest challenge to emergency response;
- It is desirable to include accident sequences evaluated in industry risk-informed EP studies in order to compare the results and insights;
- Accident sequences that have been recently analyzed in other evaluations should be selected.

Table 2-1 presents the set of accident sequences recommended for use in a regulatory oversight regimen. The letter report supporting this analysis and the Table 2-1 sequences is included in Attachment 1. The selected accidents include important risk contributors with credible frequencies. The spectrum includes accidents initiated by random plant failures, external hazards such as earthquakes, and hostile action events. All of the sequences selected result in relatively early releases. Long-term scenarios were considered but were eliminated because sufficient time would be available to complete necessary emergency response actions. Similarly, the recommended sequences would result in substantial releases of radionuclides because of either containment failure or bypass. Scenarios involving only containment leakage were not considered because they produce a small source term and, if necessary, emergency response actions could be completed during a leakage event with a lower risk to the public health and safety than the LER scenarios. Each of the selected sequences could be caused by multiple hazards or by hostile action. For some of the scenarios, hostile action was assumed to speed up the timing.

Table 2-1. Accident Sequences

Accident Scenario	Selection Criteria					
	Accident can be caused by multiple hazards?	Accident applicable to other plants?	Scenario important in recent models/ studies?	Sequence reflects early release potential?	Sequence included in industry studies?	Recent MELCOR analysis of sequence?
PWR						
Short-term station blackout (SBO), immediate loss of TDAFW, consequential SGTR	Yes	Yes	Yes	Yes	Yes	Yes
Large Loss of Coolant Accident (LOCA), failure of coolant injection, early containment failure	Yes	Yes	Yes	Yes	Yes	Yes
BWR						
Short-term SBO (with stuck open relief valve (SORV)), failure of turbine-driven systems	Yes	Yes	Yes	Yes	Yes	Yes
Interfacing System LOCA (ISLOCA)	Yes	Yes	Yes	Yes	Yes	No

The objective of this study was to determine through a proof of concept whether an analytical technique could be developed to risk inform EP oversight. To achieve the objective, the above suite of credible scenarios, important to emergency planning, were developed. These scenarios are not intended to be considered the only applicable scenarios. However, these were sufficient to demonstrate the analytical techniques developed in this study.

3.0   MELCOR ANALYSES

MELCOR is a computer code, developed by Sandia National Laboratories for the NRC, which models the progression of severe accidents in PWRs and BWRs. A broad spectrum of accident phenomena in both PWR and BWR are treated within the code. The MELCOR models used in these analyses represent a current state of knowledge in modeling for the two reference plants. Significant changes have been made during the last twenty years in the approach to modeling core behavior and core melt progression, as well as the nodalization and treatment of coolant flow. MELMACCS compiles MELCOR outputs into a radionuclide source term for transition into part of a MACCS2 (MELCOR Accident Consequence Code System Version 2) input.

3.1   Reference Site 1 Plant Model Accident Scenarios

Each of the reference Site 1 PWR plant model accident scenarios are described below.

3.1.1   Large Break LOCA with Early Containment Failure

The large break loss of coolant accident (LBLOCA) modeled is a pipe diameter size break occurring in the hot leg. It has been assumed to be caused by a hostile action, as this type of break is unlikely to occur during normal plant operation or be caused by a seismic event. The break will cause the primary coolant to leak at a very large rate into the containment, leading to the uncovering of the fuel in the reactor vessel. To access the hot leg, it is assumed that hostile forces breach the containment building. The breach is large enough to keep the containment at or near atmospheric pressure for the entire duration of the accident.

The hostile action also includes a disabling of the water supply for the emergency core cooling system (ECCS) and containment sprays. The ECCS provides an additional supply of coolant to the core in the event of an accident. It consists of a low pressure injection system, high pressure injection system, and nitrogen charged accumulators. The accumulators are passive systems and will function as expected, however there will not be any coolant injection from the other systems due to the loss of water supply. Both the ECCS and containment sprays will come on in recirculation mode but their only coolant source will be the primary system coolant inventory, which is insufficient for prolonged recirculation.

It is assumed that the entry of hostile forces into the reactor site was detected before their actions were successful. The reactor is successfully tripped immediately upon the awareness and there is one hour of decay heat removal before the pipe break. The containment breach and disabling of the water supply is assumed to occur at the same time the reactor is tripped. In reality there will probably be some delay before these actions are successfully completed, but as neither is of critical importance until after the LBLOCA, they are modeled as immediate.

At three and a half hours after the reactor trip, mitigative actions to repair the water supply to the ECCS system are successful. Both ECCS and containment sprays come on in injection mode and remain functional for the remainder of the transient. It was assumed the core damage progression would be halted at this point and any additional fission product releases to environment would be minimal, therefore the accident case was only modeled to this point.

3.1.2   Short Term SBO with Consequential SGTR

This accident sequence was analyzed previously as a separate NRC project, and all results shown stem from the report for that project. The short term SBO (STSBO) is initiated by a seismic event. The event causes a complete loss of all onsite and offsite AC and DC power, which in addition to physical damage from the seismic event, results in all active ECCS systems failing, as well as the turbine driven auxiliary feedwater (TDAFW) pumps.

Excessive cycling causes a safety relief valve in the secondary coolant side to stick open. The stuck open relief valve (SORV) leads to a thermally induced steam generator tube rupture (SGTR). In this case, two steam generator tubes rupture soon after the SORV, which opens up a bypass pathway for radionuclides to transport from the core to the environment.

3.2   Reference Site 2 Plant Model Accident Scenarios

Each of the reference Site 2 BWR plant model accident scenarios are described below.

3.2.1   Short Term SBO with Failure of Turbine-Driven Systems and SORV

This accident sequence was analyzed previously as a separate NRC project, and all results shown stem from the report for that project. An STSBO can be caused by an internal fire or flood, but the primary contributor for this analysis is a seismic event. The event causes a complete loss of all onsite and offsite AC power, including both emergency AC diesel generators and the DC batteries.

This case includes the failure of the emergency coolant makeup systems (i.e., reactor core isolation cooling (RCIC) and high pressure coolant injection (HPCI) systems which use steam-driven turbines to provide make up coolant to the reactor pressure vessel (RPV). The failure of the RCIC and HPIC systems is due to a loss of DC electric control of the systems. It is possible to manually actuate or black-start the systems; however such actions were not credited in this scenario.

3.2.2   Short Term SBO with Interfacing Systems LOCA

This accident is assumed to be initiated by a hostile action. It includes a station blackout with the same sequence as the previous scenario, but also includes a break in the reactor water clean-up (RWCU) system. The break occurs in the inlet piping to the RWCU system, outside primary containment, which is why it is referred to as interfacing system. Since the reactor building itself is not a containment structure, this break results in a bypass pathway for fission products to reach the environment. There are isolation valves on the RWCU piping which would normally be shut in the event of an accident. However, it is assumed attempts to shut the valves are not successful.

Since this event is initiated by a hostile action, early warning was considered. There are a number of early mitigative actions that would occur, but the two that are most relevant are an immediate reactor SCRAM and the isolation of the RWCU system. This means the closure of the isolation valves would occur before power is lost and there would be no LOCA. There would still be a loss of power, however the only difference between this scenario crediting early warning and the STSBO described in Section 3.2.1 would be the short time for successful decay heat removal between SCRAM and the loss of power. This was decided to have a minimal impact on the results of the accident, therefore early warning was not credited for this case.

3.3   Reference Site 1 Large Break LOCA Plant Model Results

Table 3-1 summarized the timing of the key events during the large break LOCA transient. This transient included an early warning action. The reactor SCRAM occurs immediately and is successful. The break in containment and disabling of the water supply also occurs immediately. After the LBLOCA occurs at one hour, recirculation comes on briefly but cannot be sustained, and the accumulators drain in a matter of seconds. Fission product release from the fuel begins at 2 hours 37 minutes after the LBLOCA, which is a short time when compared to most design-basis events (DBEs). Because of the immediate failure of containment, there is an early release of radionuclides to the environment.

Table 3-1. Key Events During LBLOCA

Event Description	Time (hh:mm)
Early Warning of Hostile Action: Reactor trip TCV closed AFW (motor driven and turbine driven) started Containment breached Water supply disabled	00:00
Hostile Action successful: Pipe Diameter Break in Hot Leg	01:00
Accumulators start discharging Containment sprays initiated in recirculation mode ECCS (HHSI & LHSI) initiated in recirculation mode	01:00
Accumulators are empty	01:01
Recirculation terminated due to low water mass in sump	01:10
Stuck open SG PORV	01:38
Switchover from ECST to ECMT for AFW source	02:00
Start of fuel heat-up	02:02
First fission product gap releases	02:37
Mitigative Action successful: Containment sprays initiated in injection mode ECCS (HHSI & LHSI) initiated in injection mode	03:30
End of calculation	03:30

3.3.1 Thermal-Hydraulic Results

Figure 3-1 shows the water levels in the active core and lower head of the RPV. The core level is maintained for the first hour of the transient following the successful SCRAM. At one hour the large break LOCA occurs and the water level in the RPV drops rapidly as system depressurization forces coolant out the break. This coolant flows into one of the steam generator rooms of the containment, which drains into the basement of the containment.

Recirculation pumps have been initiated, and the ECCS system pumps most of the water back into the RPV, although some goes to containment sprays. This causes the water level in the RPV to rise back to its initial value. However, the water in the sump is quickly exhausted by recirculation and the water in the RPV begins to be boiled away by decay heat, which continues for the duration of the sequence. The active core, which occupied the volume of the vessel between 3.0643 m and 6.7217 m, is essentially uncovered by the end of the modeled transient.

Fuel temperatures drop quickly following the reactor SCRAM as the reactor power decreases from full power to decay heat levels, as seen in Figure 3-2. For the remainder of the hour prior to the LBLOCA, temperatures drop slowly as the decay heat is removed by the steam

13

generators. When the LBLOCA occurs, the water in the core is quickly dumped into containment. The combination of a heat exchanger cooling the water being recirculated back into the RPV and the cold water from the accumulators creates a new heat sink which is enough to drop fuel temperatures about 150°C.

Once the fuel starts to become mostly uncovered, its temperature starts to rise dramatically. The first fission product releases due to cladding failure happen at 2 hours 37 minutes while the first debris relocation, caused by collapsed fuel, occurs at 2 hours 46 minutes. By 3 hours 12 minutes, the hottest fuel/cladding debris has reached its melting temperature of 2800 K. However, by the time coolant is injected again at 3 hours 30 minutes, there have not been any failures in the core lower support plate or the lower head. The cold water begins to cool the intact fuel and melted debris. There is no further core failure or fuel relocation after this point.

Pressure in the reactor vessel, which is initially at about 2250 psi, drops immediately to near atmospheric after the LBLOCA occurs as the high temperature and pressure water in the primary system flows into the containment. The pressure stays near atmospheric, even as the water remaining in the RPV is boiled to steam. The pressure in containment, as seen in Figure 3-3, is initially slightly below atmosphere so that any break in containment will result in a net flow into rather than out of containment. Due to code restrictions in which all time dependent volumes must become active at the same time, it was necessary to keep containment at this lower than atmospheric pressure until the LBLOCA. The large containment break which occurs immediately would quickly raise pressure to atmospheric levels, but this did not affect results as nothing significant happens outside of the RPV until after the LBLOCA. Once the LBLOCA occurs at one hour, there is a large spike in containment pressure as the water from the PRV encounters the lower pressure of containment and immediately flashes into steam. However, the size of the containment break allows for almost immediate depressurization and the containment pressure quickly returns to atmospheric pressure, where it stays for the duration of the modeled transient.

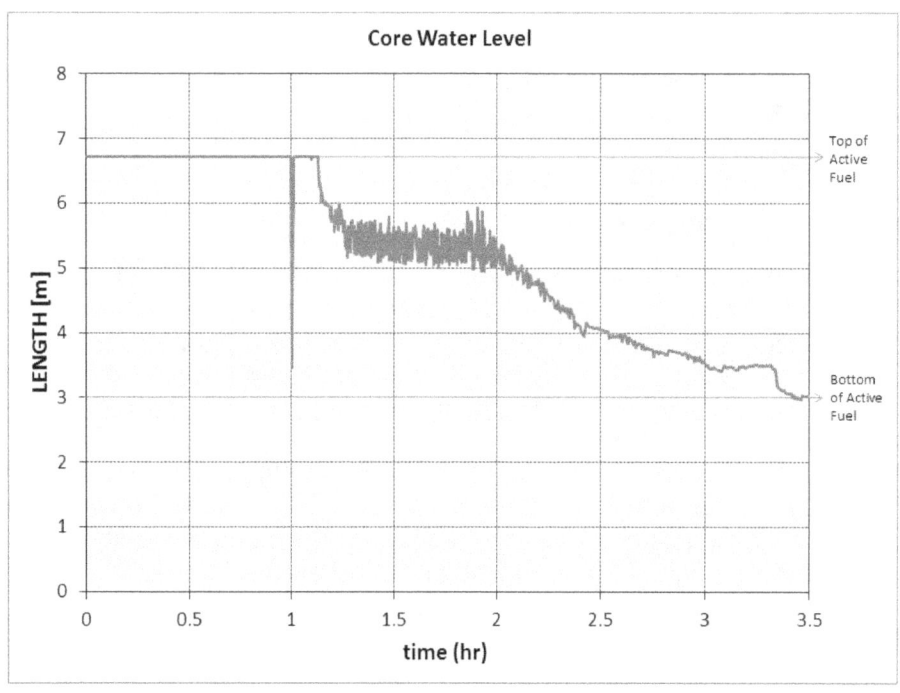

Figure 3-1. Water Level in the Active Core

14

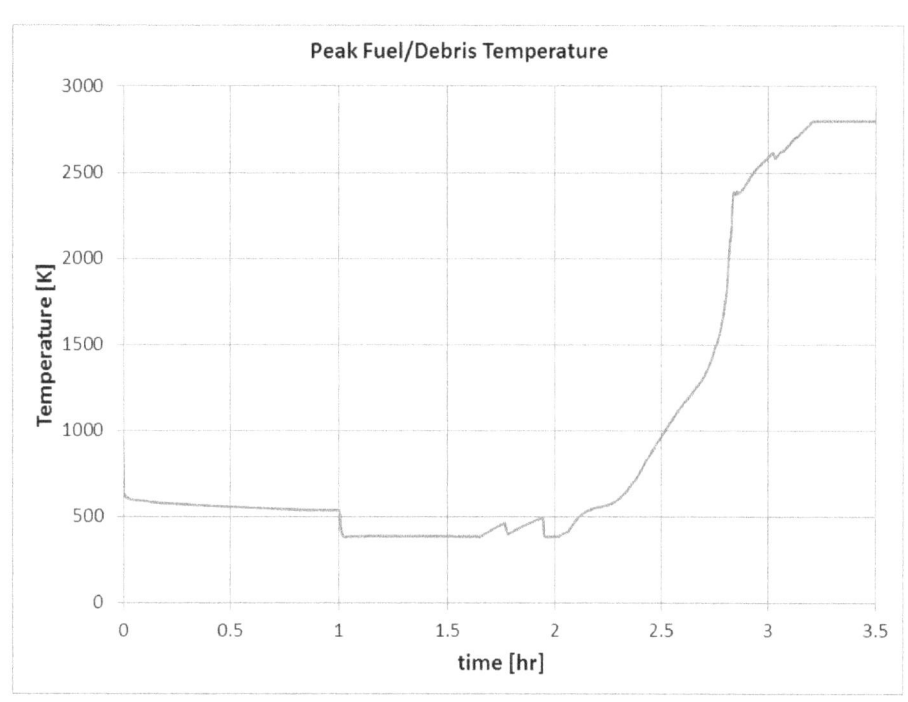

Figure 3-2. Peak Temperature of Fuel and Debris

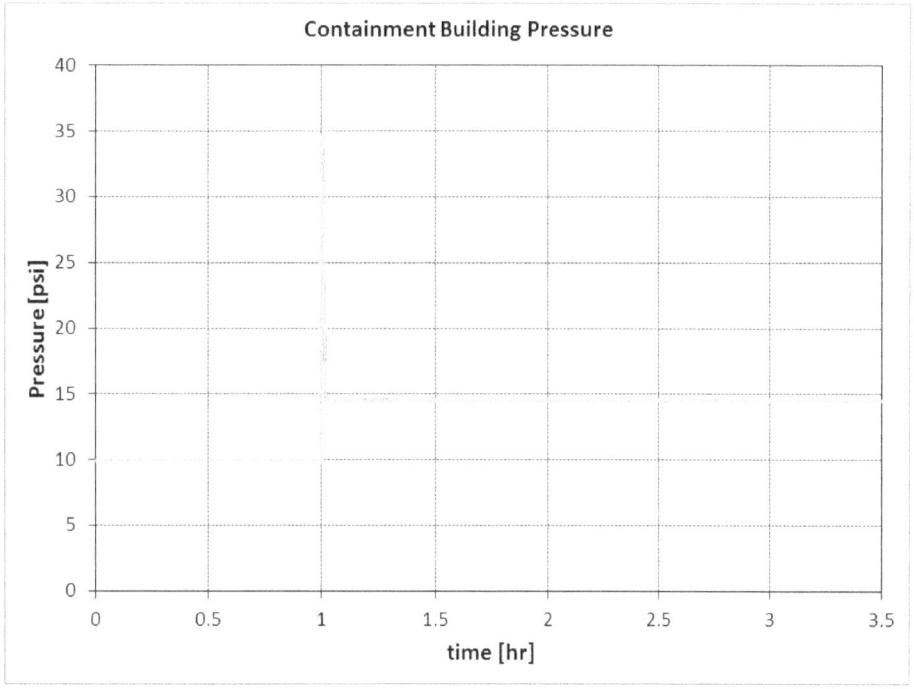

Figure 3-3. Containment Building Pressure

3.3.2 Radionuclide Results

The extreme break size of the LBLOCA causes the RPV to drain quickly, but the subsequent recirculation and boil down leads to the fuel not heating up until 1 hour 2 minutes after the

15

LBLOCA. Fission product release from fuel rod gaps begins at 2 hours 37 minutes, due to cladding failure. Because of the location of the LBLOCA, a majority of the released fission products are released into containment. In MELCOR, it is assumed that all elemental iodine immediately combines with cesium to form cesium iodide (CsI). Figure 3-4 shows the distribution of CsI in the containment, RPV and released into the environment for the duration of the transient modeled. After gap failure, the majority of released CsI is airborne in the containment, since containment sprays are not on during most of the release period. A small portion remains contained within the RPV and another small portion is deposited on the containment surfaces. A large fraction, when compared to other LWR accident calculations, is released to the environment through the large break in containment. By three and a half hours the environmental release of CsI is 8.1% of the initial inventory.

Figure 3-5 shows the fraction of the initial radionuclide inventory that is released to the environment for all relevant radionuclide classes for the duration of the modeled transient. Fission products start being released from fuel at 2 hours 37 minutes and immediately begin to be transported to the environment through the large breach in containment since containment spray is not on for most of the release period. The fact that the containment is at atmospheric pressure reduces the total release to the environment because there is no large pressure differential to force flow out the containment breach. Regardless, many of the radionuclide classes have been released in significant quantities (i.e., >1%) by the end of the modeled transient, including noble gases, iodine and cesium. At three and a half hours containment spray is restored in injection mode and remains on indefinitely through the combination of injection and recirculation, and therefore additional radionuclide release to the environment is expected to be minimal. The releases are used to create a source term used by the MACCS2 to model environmental and health effects.

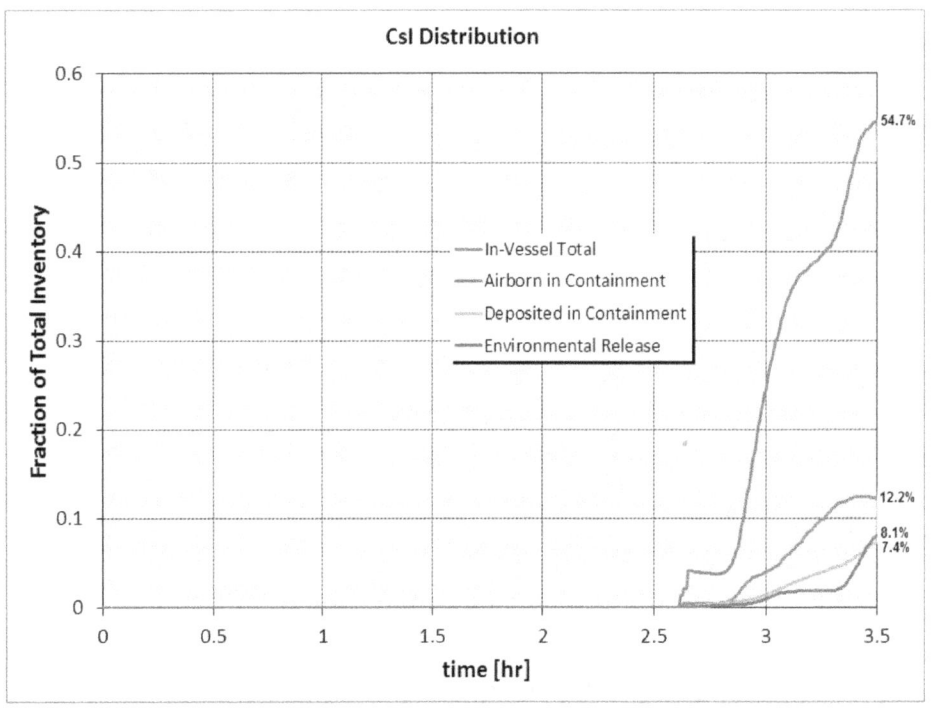

Figure 3-4. Cesium Iodide Distribution in the Containment, RPV, and Environment

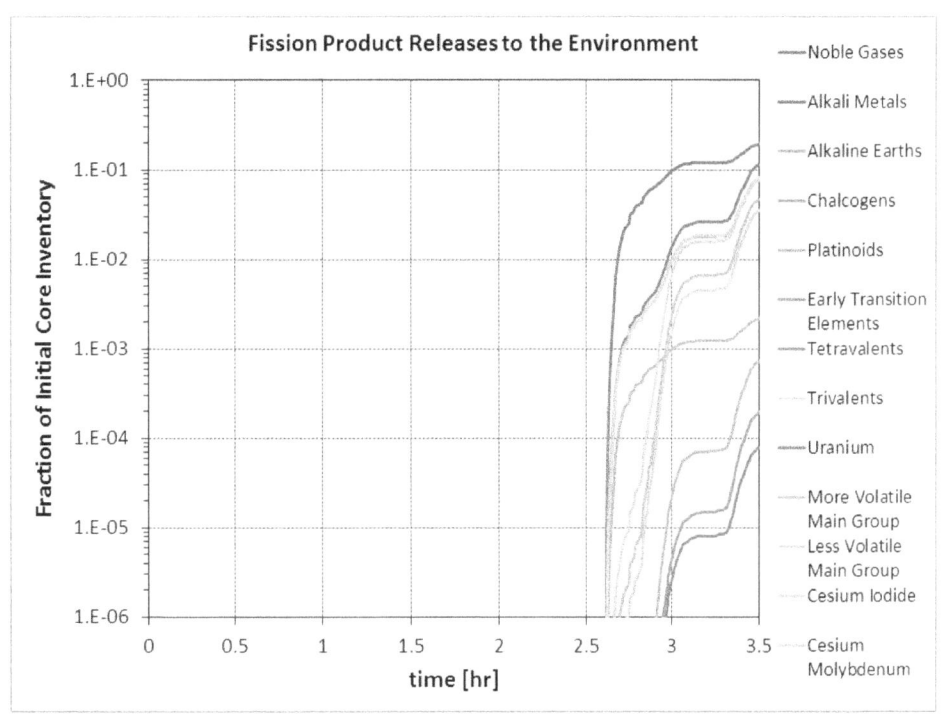

Figure 3-5. Environmental Releases for all Fission Product Groups

3.4 Reference Site 1 STSBO with Consequential SGTR Plant Model Results

Table 3-2 summarizes the timing of key events during the short term station blackout transient at the PWR. This accident included a thermally induced steam generator tube rupture, which is caused by a combination of tube heating and a pressure difference caused by a stuck open relief valve in the secondary system. Two cases were considered in the previous studies: One steam generator tube failure (100% flow area) and two tubes failing simultaneously (200%). For emergency response studies the source term from the 200% case was selected as it provided a slightly larger and slightly earlier environmental release, however the selected plots below will show data from both cases as they have been taken directly from the previous studies. Fuel failure does not occur until about three hours after the initiating event. This is a slower transient than the LBLOCA; however, the SGTR at about three and a half hours opens up a direct release path to the environment, so there is still a significant early release. Unlike the other three accidents modeled, this case was modeled to 96 hours. For consistency, only radionuclide releases that occur up to 48 hours after the initial event are used in the MACCS2 modeling.

Table 3-2. Key Events During STSBO with SGTR

Event Description	Time (hh:mm)
Initiating Event – Loss of all onsite and offsite AC and DC power	00:00
MSIVs close Reactor trip RCP seal leak at 21 gpm/pump TD-AFW fails	00:00
First SG SRV opening	00:03
SG dryout	01:14
Pressurizer SRV opens	01:27
PRT failure	01:47
Start of fuel heatup	02:19
RCP seal failures	02:46
First fission product gap releases	02:57
Stuck open SG PORV	03:00
SGTR	03:33
Creep rupture failure of the Loop C hot leg nozzle	03:49
Accumulator discharges	03:49
Accumulator empty	03:49
Vessel lower head failure by creep rupture	06:51
Debris discharge to reactor cavity	06:51
Cavity dryout	07:21
Containment at design pressure (45 psig)	13:36
Start of increased leakage of containment (P/P_{design} = 2.18)	30:14
Containment pressure stops decreasing	40:20
End of Calculation	96:00

3.4.1 Thermal-Hydraulic Results

The accident sequence begins with a successful reactor trip which includes a loss of main feedwater pumps and closing of the main steam isolation valves meaning the reactor loses its normal mechanism of heat removal. This causes coolant temperatures and pressures to rise on both the primary and secondary sides. Coolant from the RPV will naturally circulate through the steam generators, transferring heat by boiling away the secondary system inventory. Due to loss of AC and DC power, there are no systems available to provide additional feedwater to the steam generators and the steam generators will dry out at 1 hour 14 minutes into the transient. At this point heat removal from the core is inadequate and the primary system pressure begins to raise and is depressurized through a safety relief valve on the pressurizer causing a loss of primary system inventory as seen in Figure 3-6.

At 2 hours 19 minutes the water level in the vessel drops below the top of active fuel and the fuel begins to heat up. The uncovered cladding begins to oxidize which leads to cladding failure by 2 hours 57 minutes. The fuel temperature continues to steadily increase. At 2400 K molten zirconium cladding starts to degrade fuel into debris and at 2800 K it reaches the melting temperature.

18

Additional inventory is lost after the steam generator tube rupture at 3 hours 33 minutes. However, the primary system does not rapidly depressurize until a hot leg nozzle fails at 3 hours 47 minutes due to thermally-induced creep rupture. At this point accumulators will discharge, which raises water levels above the top of active fuel temporarily; however the accumulators are only able to delay complete RPV dryout for about an hour. RPV dryout occurs a little after 6 hours, and immediately after dryout the fuel/debris combination relocated to the lower plenum, and this leads to lower head failure at 6 hours 51 minutes.

The steam released from the core via the pressurizer safety relief valve is meant to condense as it flows through a submerged sparger in the pressurizer relief tank (PRT). However, in this case the capacity of the PRT is overwhelmed and fails, venting the steam directly into the containment. This steam causes a small increase in containment pressure, shown in Figure 3-7. There is a slight raise in pressure after the hot leg nozzle failure releases high pressure steam and hydrogen from the RPV into containment. However the majority of containment pressurization occurs after vessel failure. The debris is released to the reactor cavity where it quickly boils away water which has pooled and begins to ablate the concrete floor. The molten core concrete reaction, which continues for the remainder of the calculated transient, produces non-condensable gases which cause the majority of containment pressurization. Increased containment leakage, which occurs at 2.18 times the containment design pressure, starts at 30 hours 14 minutes. At 40 hours 20 minutes, the leakage from the containment to the environment both through the containment leakage and the steam generator tube rupture balance the gas generation and containment pressure levels and by 44 hours containment pressure begins to decline, a trend that continues through the duration of the modeled transient.

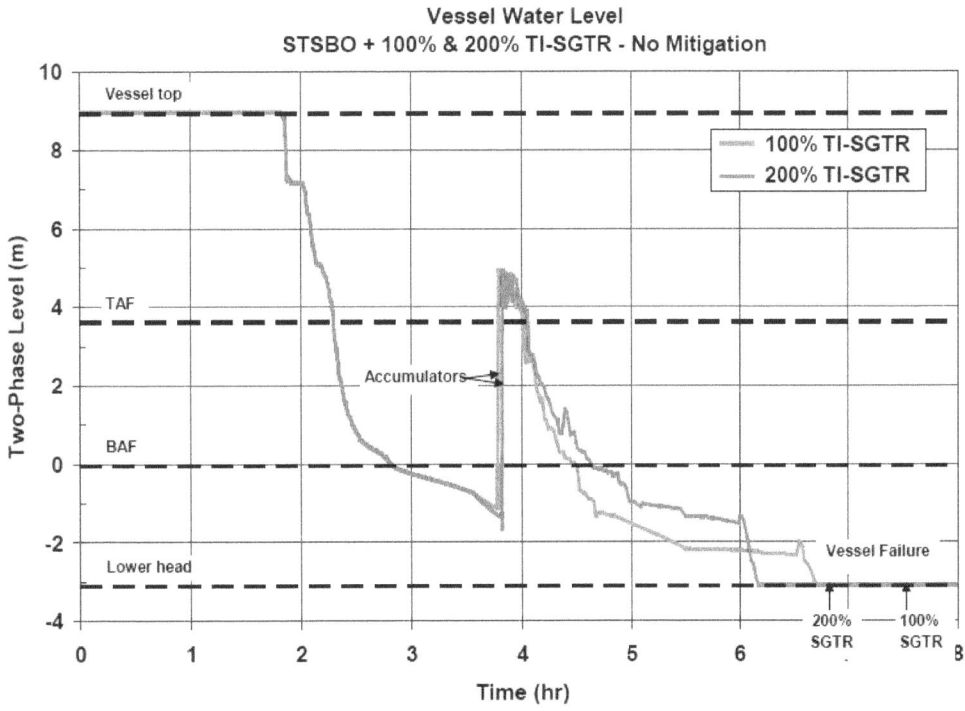

Figure 3-6. TI-SGTR STSBO Vessel Two-Phase Coolant Level

19

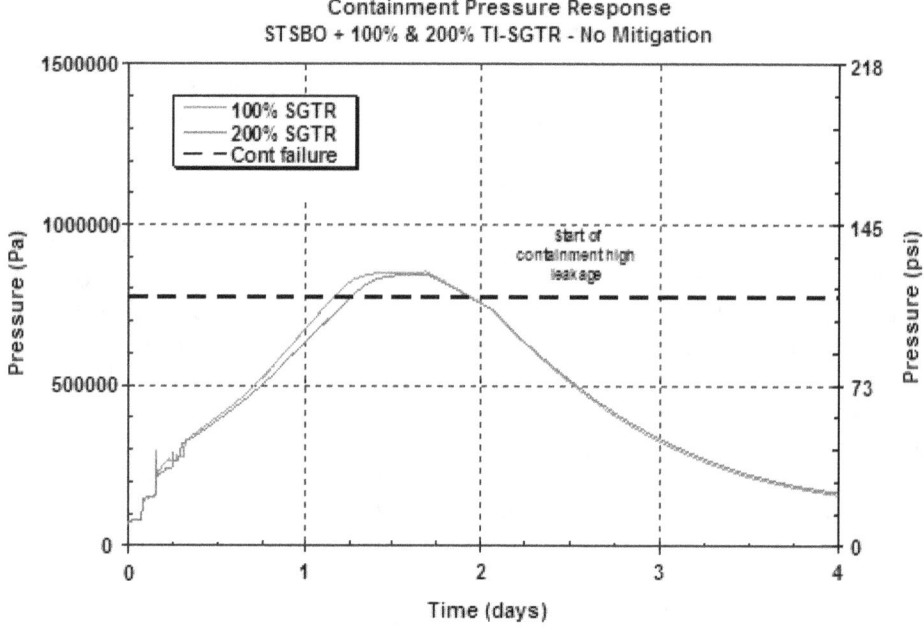

Figure 3-7. TI-SGTR STSBO Containment Pressure Histories

3.4.2 Radionuclide Results

Coolant loss from the RPV is much slower during this accident than in the LBLOCA modeled. Because of this, fuel heat-up begins about two hours later for this accident, and the first radionuclide releases do not occur until about three hours after the accident begins. Figure 3-8 shows the distribution of iodine during the accident. The first released iodine, as aerosolized cesium iodide, is contained within the RPV and primary coolant piping. According to design, as the primary coolant system pressurizes, the steam and fission products are released through a primary safety relief valve to the PRT. However, in this case, the PRT ruptures and the majority of released iodine goes from the primary safety relief valve directly into containment. The steam generator tube rupture occurs at about three and a half hours after the blackout. At that point, iodine starts to enter the secondary side of the steam generator and eventually is released to the environment at a lower amount as the steam generator is credited with a decontamination factor of seven. This release to the environment lasts for about 14 minutes since the hot leg nozzle fails and the primary system depressurizes directly to containment. There is a minimal amount of environment leakage from the containment over thirty hours after the transient begins due to over-pressurizing of the containment. However, by the end of the four days modeled, only 1.5% of the initial inventory of iodine was released to the environment. About 7% and 6% are retained in the secondary side of the steam generators and primary coolant system, respectively, while the vast majority remained inside containment and was not released.

Figure 3-9 shows the fraction of initial inventory released to the environment for all relevant radionuclide classes, including iodine. The radionuclide classes are listed by one representative element from the class. The first releases for any class occur after the SGTR. These initial releases occur as the primary coolant system slowly depressurizes through the steam generator, but then slow as the flow rate out of the steam generator drops drastically due to failure of the hot leg nozzle, which leads to rapid primary system depressurization. The next

20

significant release to environment does not occur until the increase in containment leakage occurs at about thirty hours due to containment over-pressure. These releases, for almost every radionuclide class, continue slowly but steadily until the end of the modeling period. However, at the end of the four days modeled, only the noble gases class at 95% released is released at higher than 5% of their initial inventory, with most radionuclide classes at or below 1%.

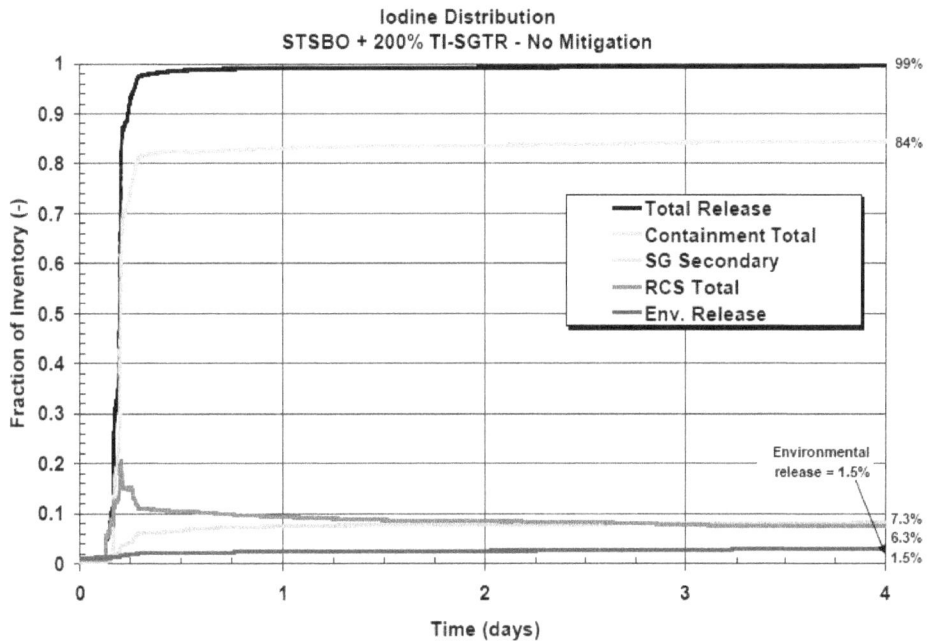

Figure 3-8. TI-SGTR STSBO Iodine Distribution

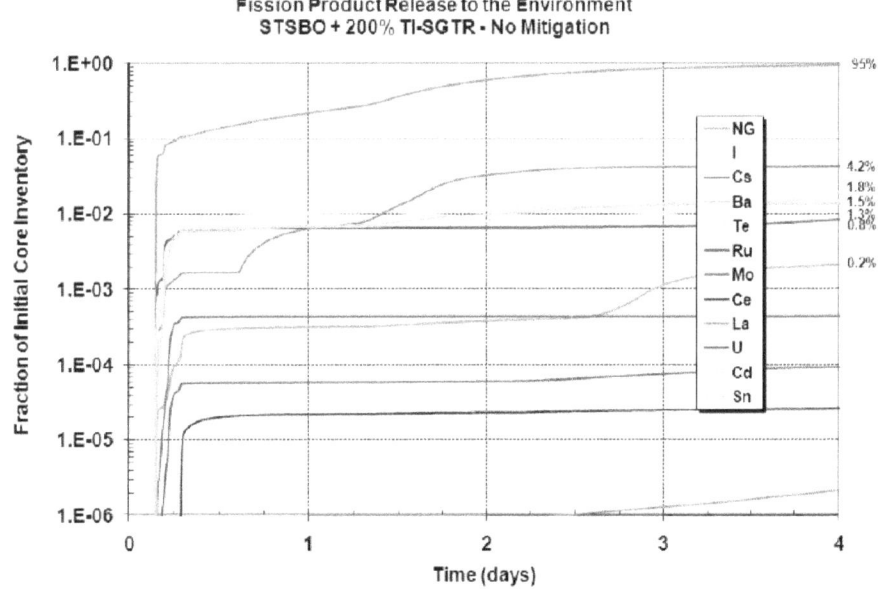

Figure 3-9. TI-SGTR STSBO Environmental Release of All Fission Products

3.5 Reference Site 2 STSBO Plant Model Results

Table 3-3 summarizes the timing of key events during the short term station blackout transient at the BWR. This accident included a stuck open relief valve between the RPV and the suppression pool, caused by excessive cycling, as well as the failure of all turbine driven injection systems (i.e., RCIC and HPCI). Fuel failure begins to occur about one hour after the initiating event, quicker than the STSBO modeled for a PWR. However, the containment (i.e., wetwell and drywell) remains intact until about eight hours after the transient begins. This sequence has the least significant environmental radionuclide releases of the four cases considered. Selected plots shown below have been taken from the NRC studies generating these results.

Table 3-3. Key Events During STSBO

Event Description	Time (hh:mm)
Initiating Event – Loss of all onsite and offsite AC and DC power	00:00
Low-level 2 and RCIC actuation signal	00:10
Downcomer water level reaches top of active fuel	00:30
First hydrogen production	01:00
First fuel cladding gap release	01:00
First channel box failure	01:12
Reactor vessel water level reaches bottom of lower core plate	02:00
SRV sticks open due to excessive cycling	02:00
RPV pressure decreases below LPI set point (400 psi)	02:18
First core support plate localized failure in supporting debris	02:36
Lower head dries out	03:30
Ring 5 CRGT Column Collapse [failed at axial level 2]	05:30
Ring 3 CRGT Column Collapse [failed at axial level 2]	05:48
Ring 1 CRGT Column Collapse [failed at axial level 1]	05:54
Ring 4 CRGT Column Collapse [failed at axial level 1]	06:06
Ring 2 CRGT Column Collapse [failed at axial level 1]	06:06
Lower head failure (yield from creep rupture)	07:54
Drywell liner melt-through (leakage into torus room of reactor building)	08:12
Refueling bay to environment blowout panels open	08:12
Hydrogen burns initiated in torus room (basement) of reactor building	08:12
Door to environment through railroad access opens from overpressure	08:12
Blowout panels from RB steam tunnel to turbine building open	08:12
Steel roof of reactor building fails due to over-pressure	08:24
Reactor Pedestal through-wall erosion	11:06
Calculation terminated	48:00

3.5.1 Thermal-Hydraulic Results

The accident sequence begins with a loss of all AC and DC power along with a successful reactor trip, which includes an isolation of the reactor coolant system. Because of this isolation, reactor pressure quickly raises to the set point of safety relief valves (SRVs) which open to allow steam into the suppression pool and then close to maintain pressure. The SRVs cycle frequently to prevent over pressurization, and given no injection, the water level in the reactor rapidly decreases as seen in Figure 3-10.

At 30 minutes into the transient, the water level drops below the top of the active fuel and shortly after the temperatures of the fuel and cladding (Figure 3-11) start to rise. The first fuel failures due to fuel-cladding interactions start to occur at about 1 hour 30 minutes. At 2 hours, an SRV sticks open due to excessive cycling and provides a continuous pathway for steam to leave the RPV. By 4 hours, there is essentially no water left in the RPV, while slightly after 6 hours all fuel has turned to debris or melted and relocated to the lower head.

Containment pressure in the wetwell and drywell can be seen in Figure 3-12. Following the initiation of the transient, steam is immediately released to the suppression pool; however, pressure does not start to increase until the suppression pool reaches saturation. This pressure increase becomes more rapid after the SRV sticks open as steam is continually released into containment. The pressure increase slows as the RPV dries out and the lower head fails at 7 hours 54 minutes. At this point debris spreads across the drywell floor and reacts with concrete to create non-condensable gases that cause a large spike in containment pressure. However, this spike only lasts for a matter of minutes until the molten debris has breached the containment steel liner to open a pathway to the basement of the reactor building which allows a rapid depressurization of the containment. At this point, significant radionuclide release to environment begins since the reactor building is purely a support building and not a containment structure.

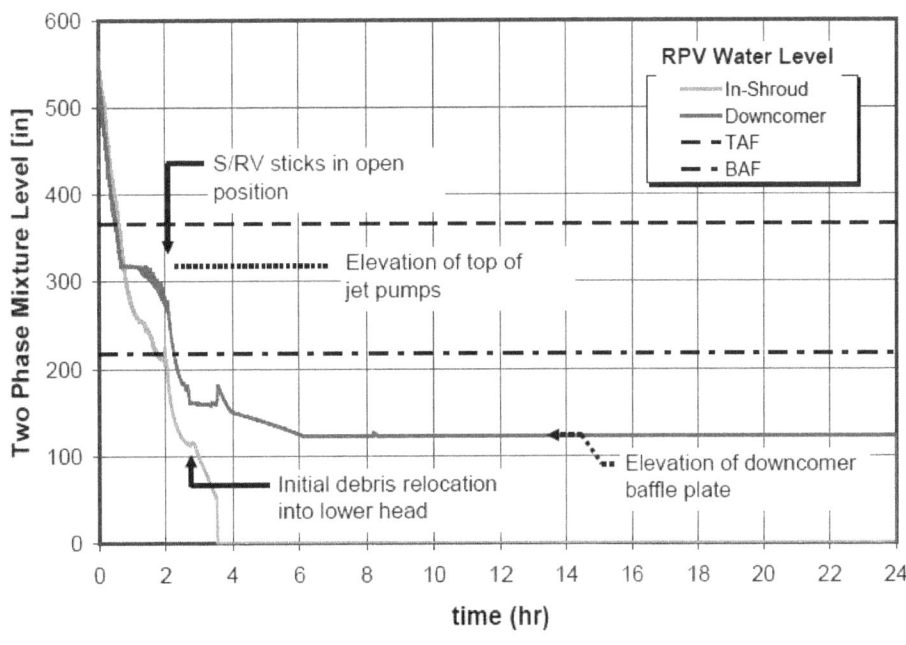

Figure 3-10. STSBO Reactor Vessel Water Level

23

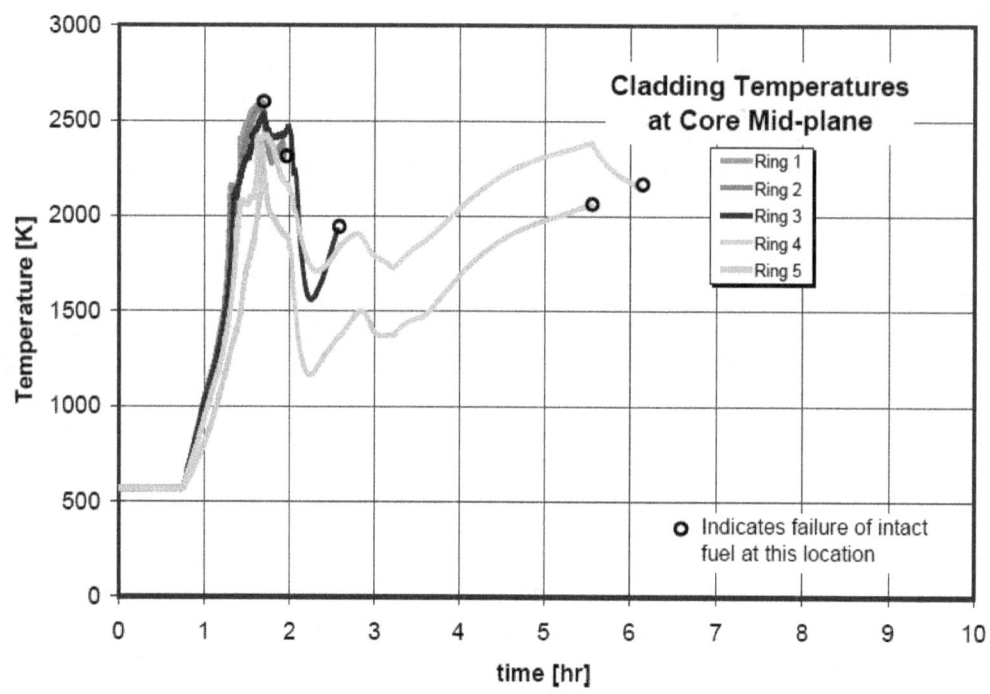

Figure 3-11. STSBO Fuel Cladding Temperatures at Core Mid-plane

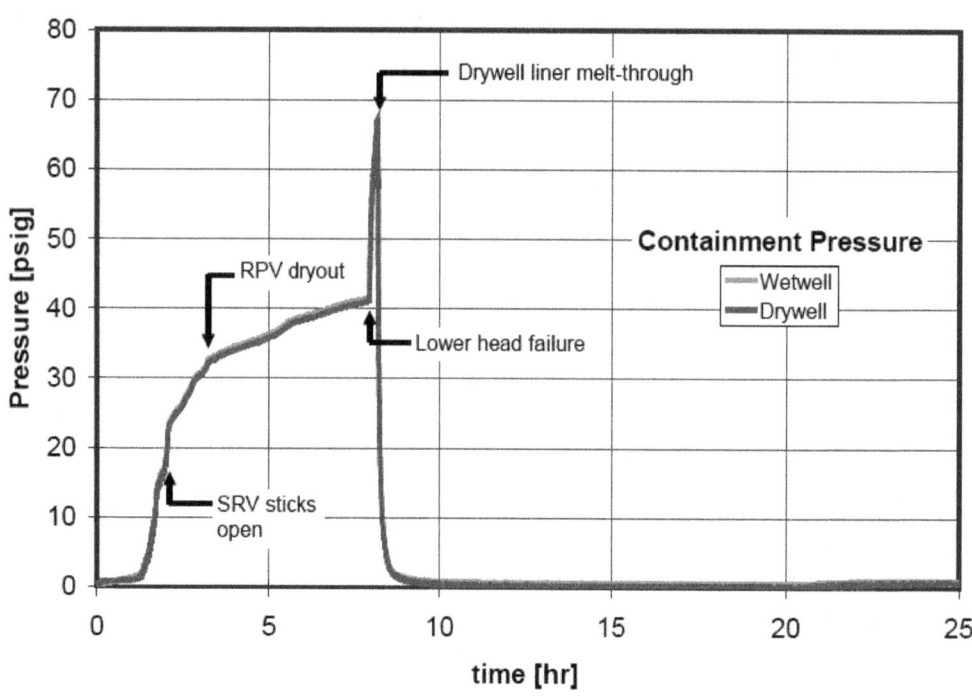

Figure 3-12. STSBO Containment Pressure History

24

3.5.2 Radionuclide Results

Due to the presence of water in the RPV up to the level of the fuel for the first thirty minutes of the transient, the first radionuclide release from the fuel does not occur until an hour after the start of the accident. Figure 3-13 shows the distribution of iodine in the RPV, containment, reactor building, and release to the environment. At first, the majority of iodine released from the fuel as aerosolized cesium iodide is airborne in the RPV, while a smaller portion is deposited on RPV surfaces. The SRV allows for the transportation of some airborne iodine into the suppression pool. When the SRV fails open at two hours, there is a large influx of iodine into the suppression pool as the RPV quickly depressurizes. When the RPV lower head and drywall fail at about eight hours, the blowout panels in the reactor building open due to hydrogen deflagrations, and iodine release to the environment begins. The iodine within the suppression pool is contained; however most of the iodine still within the RPV is gradually released to the environment. A small portion is retained by the reactor building. By the end of the 48 hour modeled transient, about 10% of the initial inventory of iodine has been released to the environment.

Figure 3-14 shows the fraction of initial inventory released to the environment for all relevant radionuclide classes. The radionuclide classes are listed by one representative element from the class. There are no releases for any radionuclide class until the lower head and subsequently the drywell liner failure by melt through at about eight hours. At this point there is a large puff release as the containment rapidly depressurizes to the reactor building and ultimately to the environment. This initial release accounts for the majority of the release for most radionuclide classes. While there are some additional releases during the remainder of the modeled transient, at its completion no radionuclide class has released more than 10% of its initial inventory, except noble gases, which have essentially a 100% environmental release.

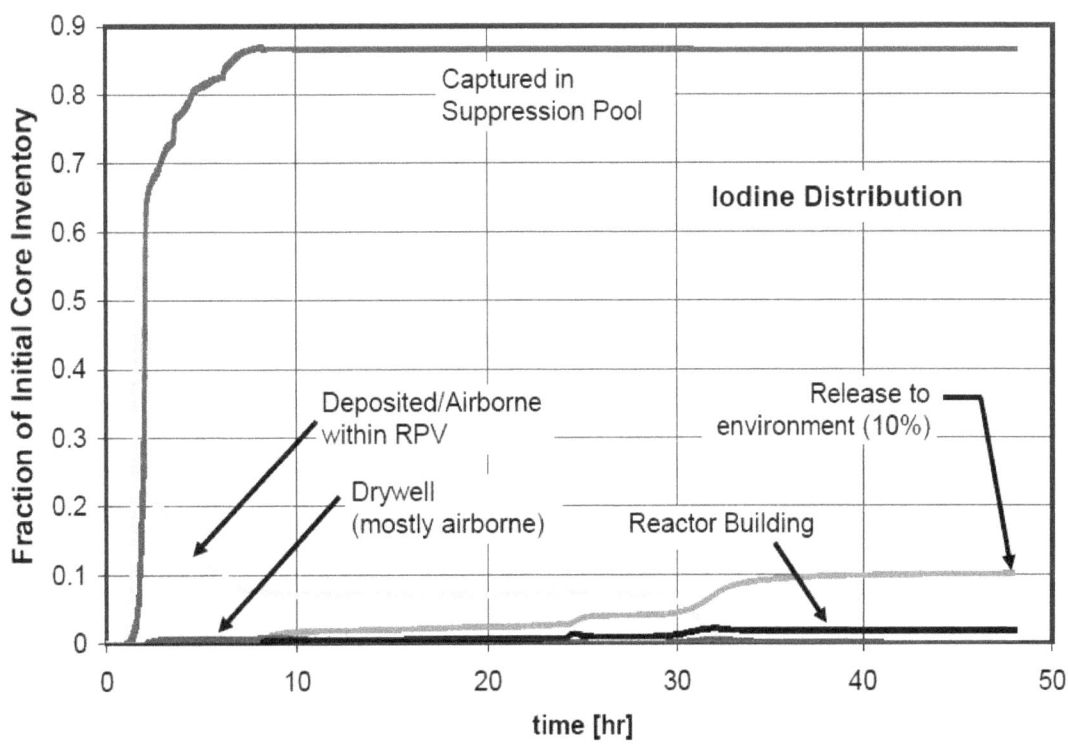

Figure 3-13. STSBO Iodine Fission Product Distribution

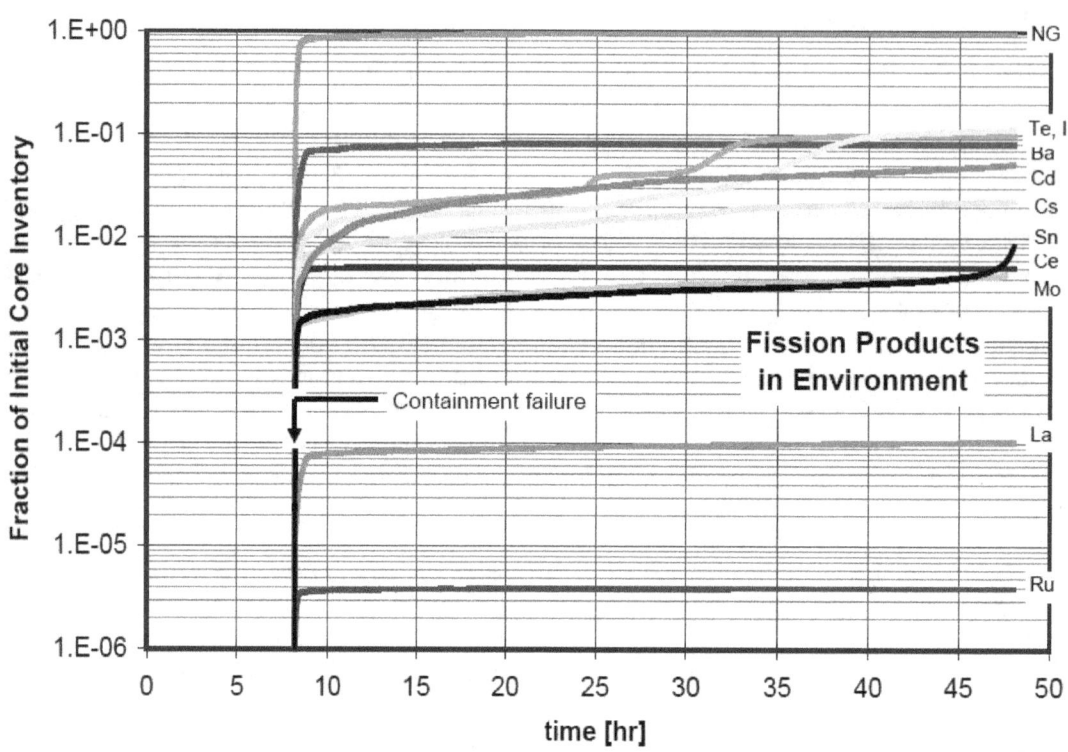

Figure 3-14. STSBO Environmental Source Term

3.6 Reference Site 2 STSBO with Interfacing Systems LOCA Plant Model Results

Table 3-4 summarizes the timing of key events during the short term station blackout transient with interfacing systems LOCA (ISLOCA) at the BWR. The LOCA occurs in the reactor water clean-up system and releases coolant directly from the RPV into the reactor building, which is not a sealed containment. Fuel failure begins to occur only fourteen minutes after the initiating event, much quicker than the case which was only an STSBO. Although the primary containment does not fail until 5 hours 35 minutes into the accident, the location of the ISLOCA leads to the quickest significant environmental releases of radionuclides for the four cases considered.

Table 3-4. Key Events During STSBO with ISLOCA

Event Description	Time (hh:mm)
Initiating Event – Loss of all onsite and offsite AC and DC power with break in reactor water clean-up system	00:00
Refueling bay to environment blowout panels open	00:00
Downcomer water level reaches top of active fuel	00:01
First hydrogen production	00:02
RPV pressure decreases below LPI set point (400 psi)	00:09
First fuel-cladding gap release	00:14
Reactor vessel water level reaches bottom of lower core plate	00:14
First channel box failure	00:22

Event Description	Time (hh:mm)
First particulate debris created by collapsing fuel	00:22
First hydrogen burns initiated in reactor building	00:29
Door to environment through railroad access opens from overpressure	00:30
Blowout panels from RB steam tunnel to turbine building open	00:30
First core support plate localized failure in supporting debris	00:33
Lower head dries out	01:45
Lower head failure from thru-wall yielding	05:21
Drywell liner melt-through (leakage into torus room of reactor building)	05:35
Reactor Pedestal through-wall erosion	07:29
Calculation terminated	48:00

3.6.1 Thermal-Hydraulic Results

The accident sequence includes a loss of all AC and DC power along with a double ended pipe break outside containment. The reactor trip is successful; however, the isolation of the break is not, allowing water from the RPV to flow out the pipe break into the reactor building. Due to the size of the break, water levels in the RPV (i.e., see Figure 3-15) drop very rapidly. Within 2 minutes the top of the active fuel is exposed and within 15 minutes the water level has dropped below the bottom of the active fuel. The pressure vessel is essentially dried out within 2 hours of the accident initiation.

The uncovering of the fuel is accompanied by an immediate increase in fuel temperature (i.e., see Figure 3-16). The first gap releases by cladding failure occurs at 14 minutes, while the first fuel debris formation occurs at 22 minutes. All fuel has either melted or degraded into debris at 3 hours 55 minutes into the transient. Lower head failure occurs at 5 hours 21 minutes and is accompanied by a large spike in containment pressure as seen in Figure 3-17. There is not significant pressurization of the wetwell or drywell up to this point since all water from the RPV was released into the reactor building through the RWCU pipe break, and thus no steam was released into the suppression pool. The pressure spike lasts for 14 minutes until the drywell liner experiences melt-through from coming into contact with molten corium, and molten debris is ejected into the torus room (i.e., reactor building basement), allowing rapid depressurization of containment. At this point the molten debris is cooled significantly by the water that has pooled in the torus room. There are no significant thermal-hydraulic changes for the remainder of the transient.

27

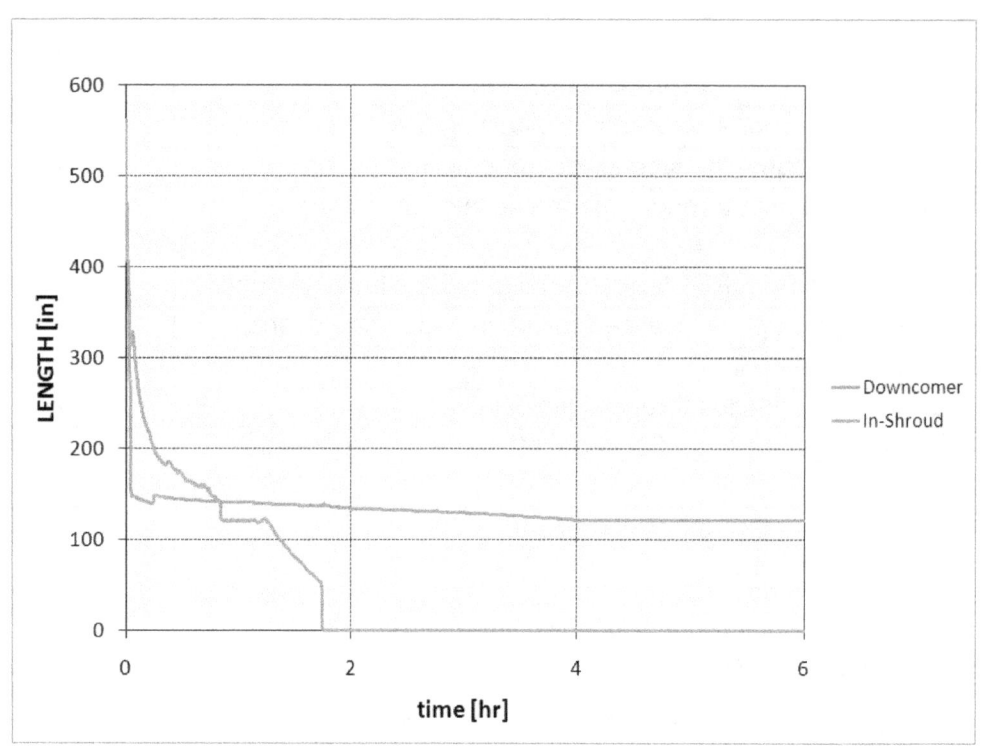

Figure 3-15. 2-Phase Water Level Inside RPV

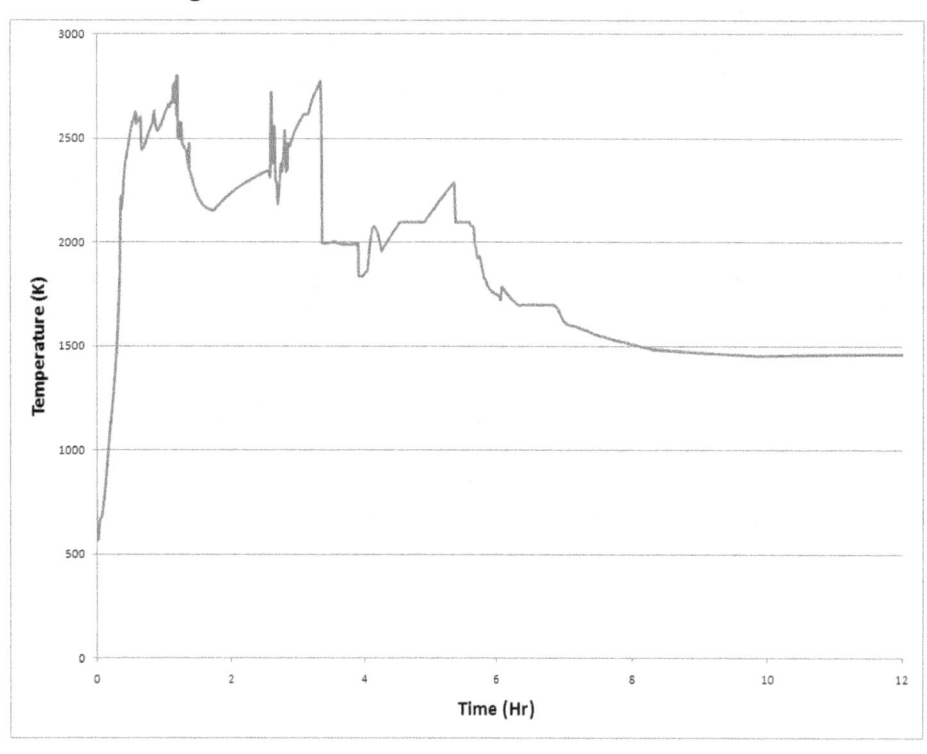

Figure 3-16. Peak Fuel/Debris Temperature

28

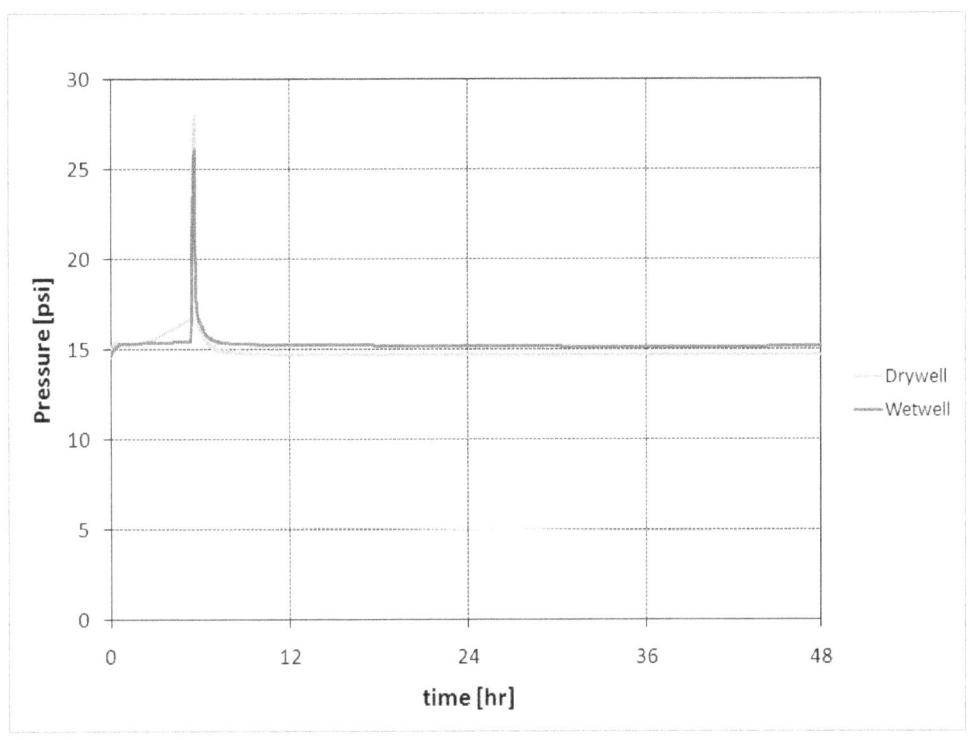

Figure 3-17. Containment Pressure

3.6.2 Radionuclide Results

Due to the size of the piping of the RWCU system where the ISLOCA occurs and the high pressure of coolant in the RPV, the fuel becomes uncovered only a minute after the transient is initiated and quickly begins to heat up. The first radionuclide releases from the fuel occur 14 minutes into the transient. Figure 3-18 shows the distribution of iodine in the containment, reactor building, and released to the environment. Between 14 and 30 minutes all iodine released from the fuel, as aerosolized cesium iodide, is contained within the reactor building. At this point, the increasing pressure in the reactor building causes the blowout panels to open, providing a release path to the environment. After this point, most of the iodine released from the fuel is released to the environment. At the end of the 48-hour transient modeled, 86.1% of the initial inventory of iodine is released to the environment, while 13% remains in the reactor building, most likely either trapped in pools or deposited on the walls and floors. A very small fraction is contained in the radwaste and turbine buildings, while an even smaller fraction is retained in the drywell and the wetwell.

Figure 3-19 shows the fraction of the initial inventory that has been released to the environment for all relevant radionuclide classes. The radionuclide classes are listed by one representative element from the class. As was shown for iodine, none of the radionuclide classes have a significant release until the blowout panels open at about thirty minutes. At this point, most of the radionuclide classes have a very large release as the high pressure in the reactor building is released (i.e., puff release). Two classes (Ba and Ce) do not have significant releases until after the molten core melts though the drywell at about five and a half hours. By the end of the modeled transient two of the classes (Ru and La) have less than 1% of the total inventory released to the environment, while only 3.3% of the Ce class is released. All other relevant radionuclide classes have more than 20% of the initial inventory released, with the majority of these radionuclide classes above 50%.

29

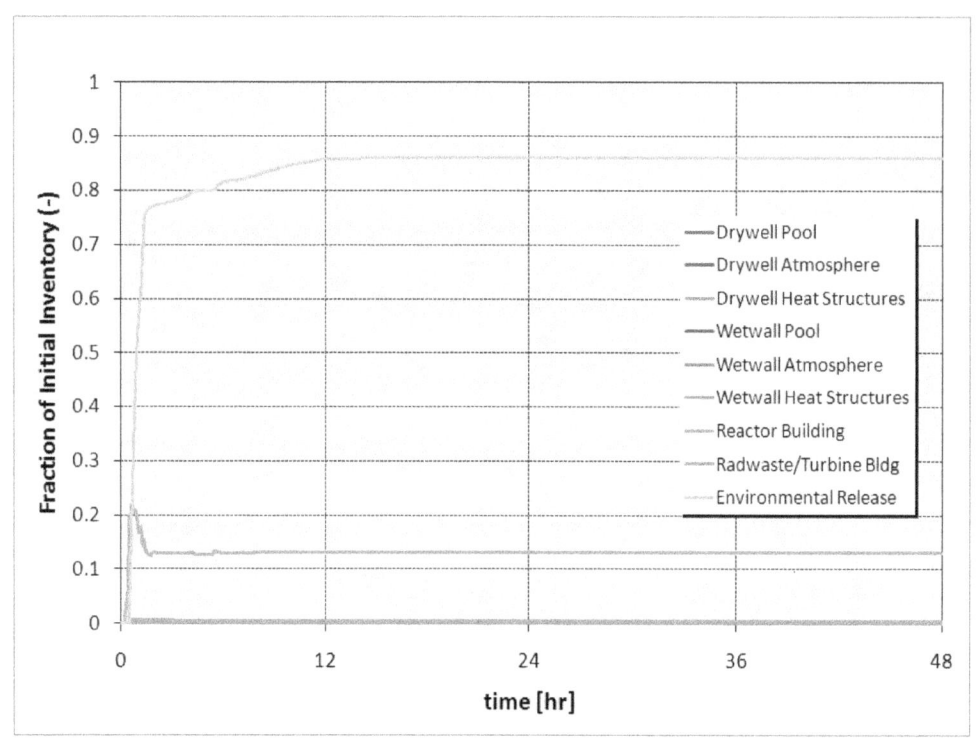

Figure 3-18. Iodine Distribution

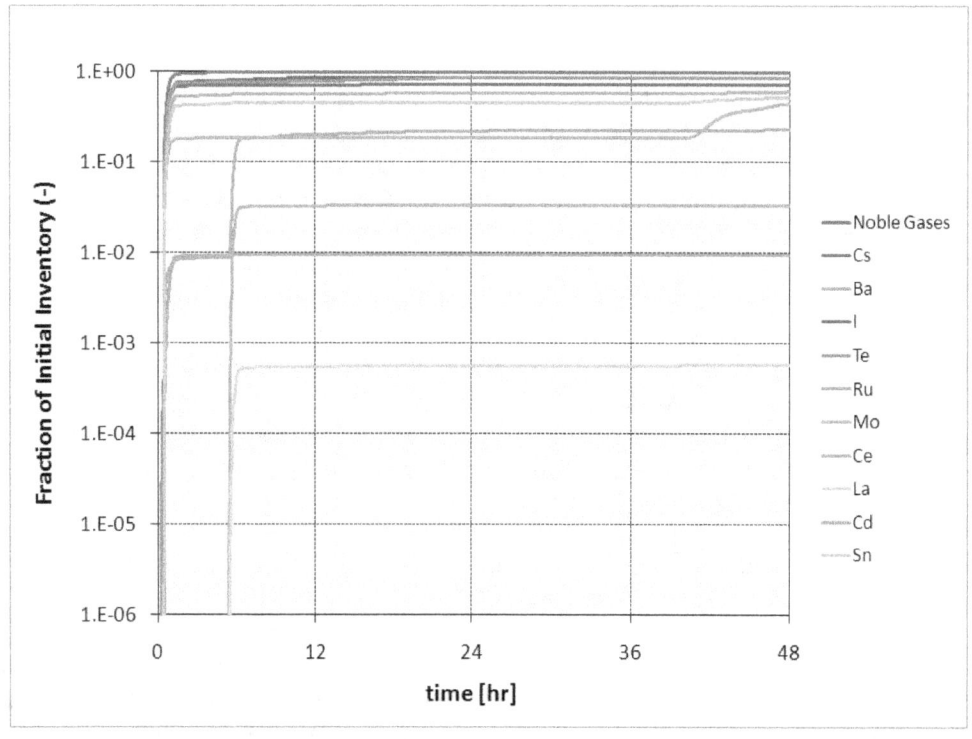

Figure 3-19. Environment Releases for Significant Radionuclide Classes

4.0 CONSEQUENCE MODELING

A method was developed to quantify the value of EP by calculating the difference in cumulative population dose between an ad hoc response and a response, which would follow Supplement 3 guidance. Using information specific to a site, the method applies reasoning in the development of response parameters which are input into the consequence model. Conclusions for each analysis are derived by interpreting the modeling results. The results are quantified, providing an indicator for comparison to baseline analyses results. This method is called the **De**d**U**ctive **Q**uantification **I**ndex (DUQI) method, the foundation of which is the consequence analysis.

Consequence analysis modeling was performed for the two reference sites, two accident sequences, and the specific scenarios identified in Figure 4-1 which are described below:

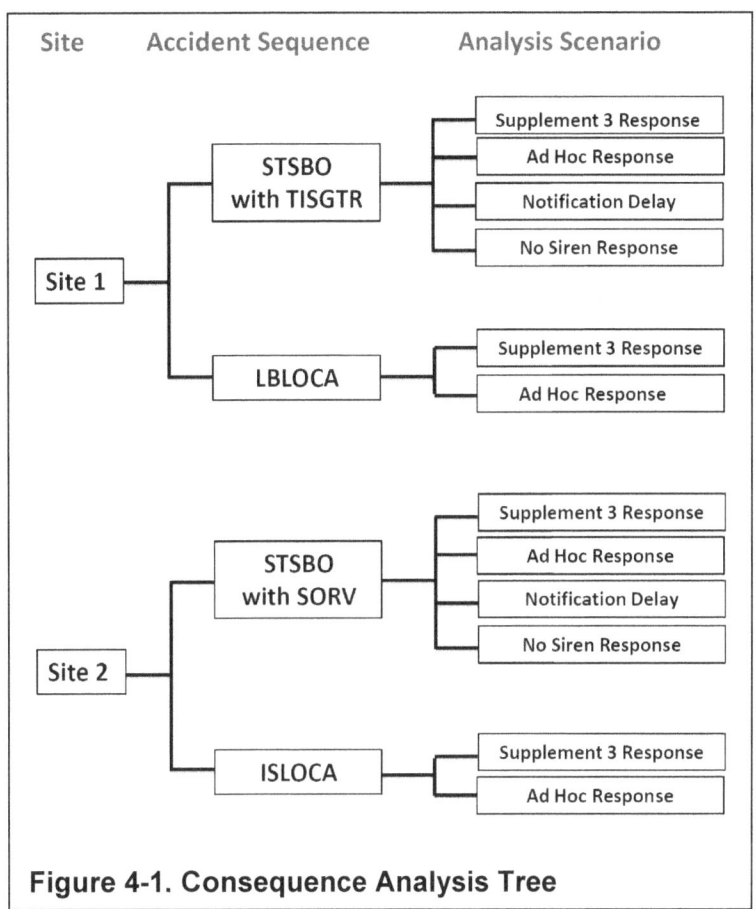

Figure 4-1. Consequence Analysis Tree

- Supplement 3 Response: This is the baseline analysis following the PAR strategy identified in the update to Supplement 3 of NUREG-0654/FEMA-REP-1, Rev. 1 (NRC, 2011a) and includes the application of staged evacuation;
- Ad Hoc Response: This analysis assumes there is no formal EP program established to respond to NPP accident conditions. It is assumed that OROs are in place and have the standard training and resources to respond to all hazards emergency plans;
- EP Element Analysis is an evaluation of a response that assumes sirens are not operable in the 2-5 mile area around the NPP;
- Notification Delay is an evaluation of a response assuming that there is a 1-hour delay in the initial notification from the plant to OROs.

The major components of the DUQI method include: 1) Baseline Analysis; 2) EP Parameter Analysis; and 3) Quantification. The metric used in this demonstration project was cumulative population dose which was estimated using the MELCOR Accident Consequence Code System Version 2 (MACCS2) model. This project utilized the most recent version of the MACCS2 analysis code and the MACCS2 graphical user interface called WinMACCS, which employs code enhancements that simplify user input, improve code performance, and enhance existing functionality.

The MACCS2 code integrates four modules that include ATMOS, EARLY, CHRONC, and COMIDA (NRC, 1998b). For this analysis, only the ATMOS and EARLY modules of MACCS2 were used. ATMOS was used for atmospheric transport and deposition, and EARLY was used to perform the emergency phase calculations. CHRONC is used for intermediate and long term phase calculations, which were not needed to support early and near field effects. COMIDA is the ingestion pathway model, also not needed to support early and near field effects. The early phase calculations assumed an emergency phase duration of 1 week (7 days), which gave sufficient time for the plume to exit the problem domain. The 95th percentile cumulative population dose results were used to support the study conclusions.

WinMACCS allows for discrete analysis of individual segments of the population by establishing cohorts. The user is able to identify multiple cohorts, each of which represent a segment of the population that has different response characteristics than other population segments. The number of cohorts is not limited, but there is diminishing value in establishing a large number of cohorts because the response characteristics begin to overlap within the evacuation period and the effects on different cohorts become indistinguishable. In this study, the general public was separated into 5 discrete cohorts for each reference site to allow a dispersed loading of the public onto the roadway network. This allowed improved simulation of evacuation road network loading. A total of 12 cohorts were established for Site 1 and 11 cohorts were established for Site 2. Establishing this number of cohorts allowed simulation of large transient facilities, such as amusement parks, to be modeled in a summer scenario for Site 1 and allowed simulation of a winter scenario that included schools evacuating for Site 2. The large number of cohorts and the approach to modeling for this project represents the highest fidelity use of the MACCS2 modeling code ever attempted.

4.1   Population Cohorts

Site 1 represents a high population density site. This site has a large summertime transient population that includes high attendance attractions. There is a large transient employee population that commutes into the EPZ during the day to work. Because this site has a large summertime transient population, a summer scenario was developed. Twelve cohorts were established for this site.

Site 2 represents a moderate population density site. This site has no unique transient characteristics. A winter scenario was developed for this site and considers that schools are in session. Eleven cohorts were established for this site.

The following cohorts were common to both sites:

Cohort 1 represents a shadow evacuation of 20 percent of the general public residing in the area 5 miles beyond the EPZ. A shadow evacuation occurs when members of the public evacuate from areas that are not under official evacuation orders. These generally begin when a large scale evacuation is ordered. The 20 percent estimate was derived from a national telephone survey of residents of EPZs asked questions about evacuation and protective actions (NRC, 2008).

In an evacuation, the general public will mobilize and evacuate over a period of time (Wolshon, 2010). Prior to the alert and notification of the emergency, the general public is assumed to be performing normal activities prior to evacuation (e.g., working, errands, at home, etc.). The evacuation time period therefore depends upon when they receive the warning, where they are when they receive the warning and the actions they need to take to evacuate once they understand that is the protective action order. To represent the movement of the general public over a period of time, cohorts 2 through 6 have been established as described below.

Cohort 2 represents the general public who evacuate promptly upon receiving notification and include people at home, or within the EPZ that do not return home prior to evacuating. Approximately 10 percent of the general public is assumed to mobilize and begin evacuating within 30 minutes of notification.

Cohorts 3, 4 and 5 each represent 26.6 percent of the general public. These cohorts are modeled as evacuating sequentially beginning immediately following the prompt evacuees. The cohorts were established to allow segmented roadway loading simulating the time for residents to prepare to evacuate and enter the roadway network.

Cohort 6 represents the last 10 percent of the general public to evacuate. This last 10 percent is referred to as the evacuation tail (Wolshon, 2010). The evacuation tail takes longer to evacuate for valid reasons, such as shutting down farming or manufacturing operations, performing other time consuming actions prior to evacuating, or they may have missed the initial notification.

Figure 4-2 illustrates an evacuation curve representing evacuation of the general public. This illustration is consistent with research (Wolshon, 2010) that shows a small portion of the public evacuates early and the last 10 percent of the population, referred to as the evacuation tail, takes a lengthy and disproportionate time to evacuate.

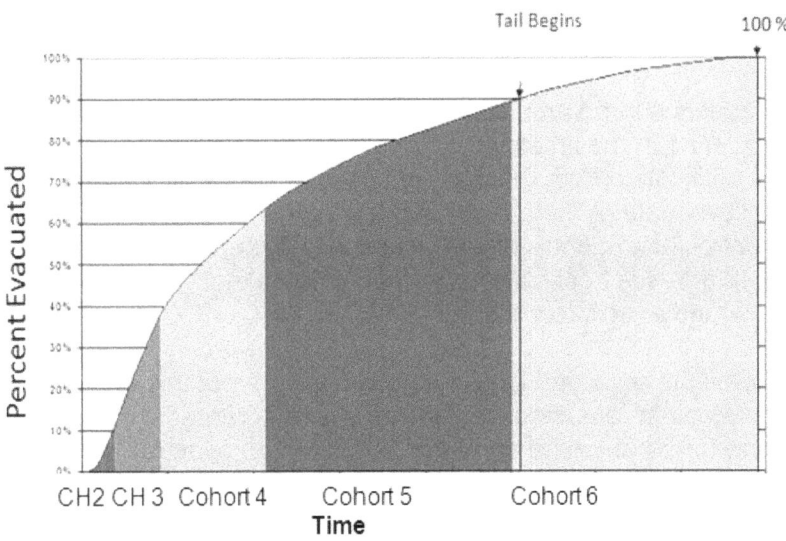

Figure 4-2. General Public Loading Curve

Cohort 7 represents the special facilities population within the EPZ which includes residents of hospitals, nursing homes, assisted living communities and prisons. These facilities are typically large and robust, providing better shielding than typical residential housing. In an emergency, Special Facilities would be evacuated individually over a period of time based upon available transportation and the number of return trips needed to evacuate a facility. As described earlier, the consequence model does not accept such input over a period of time. Because the percent of population of this cohort is very small with regard to the total population and the other cohorts, it was not necessary to separate the special facilities into multiple cohorts as was done with the general public. It was determined that an appropriate representation of this cohort in the modeling would be to start the evacuation of this cohort later in the event and apply shielding factors consistent with the types of structures within which these residents reside.

Cohort 8 represents special needs residents within the EPZ who do not reside in special facilities. Results of a national telephone survey of EPZ residents show that 6 percent (± 3.5 percent at the 95 percent confidence level) of the EPZ population may be special needs residents who do not reside in special facilities, and who would need additional assistance from outside the home in order to evacuate (NRC, 2008). Actual survey results showed 8 percent; however, a quarter of these people believed that, if necessary, they might be able to evacuate on their own.

The non-evacuating public from within the EPZ is represented as Cohort 12 for Site 1 and Cohort 11 for Site 2. This cohort represents a portion of the public who may refuse to evacuate and is assumed to be 0.5 percent of the population. Research of large scale evacuations has shown that a small percent of the public refuses to evacuate and this cohort accounts for this group (NRC, 2005a). This cohort, having decided not to evacuate, is assumed to be performing normal activities.

The above cohorts were common to both sites. Additional cohorts specific to the sites and scenarios were also developed. For Site 1, the transient population within the EPZ was divided into 3 groups. There are 2 facilities that attract large numbers of transients (Cohorts 9 and 10) and the remaining transients are distributed throughout the EPZ (Cohort 11). Three groups of transients have been established. It is assumed that some of these transients will return to their hotels to pack before evacuating the EPZ.

Site 1 Cohort 9 represents a large area tourist attraction that covers a few hundred acres represented as Transient 1 in the timelines. The transients from this facility would hear sirens and would receive a notification from the facility. Then they would complete their activities, walk to their vehicles, and evacuate. Although this attraction covers a large area, there is no preplanned traffic control for exit from this attraction. It is assumed that after hearing the siren, this cohort would wait for a site notification and then walk to their vehicles, drive to their hotel, pack their belongings, and evacuate the EPZ.

Site 1 Cohort 10 represents a second large tourist attraction, but this attraction is more concentrated (e.g., a stadium, amusement park, etc.) and is represented as Transient 2 in the timelines. The parking facility is onsite and upon receiving an evacuation order from park management, this group should be able to readily access their vehicles and evacuate the area. Visitors would walk to their nearby vehicles, drive to their hotel, pack their belongings, and evacuate the EPZ. There is no preplanned traffic control for exit from this attraction.

Site 1 Cohort 11 represents the remaining transients in the area including employees who work within the EPZ but do not live within the EPZ, including visitors, shoppers, etc. This group is dispersed throughout the EPZ and receives the warning generally at the same time as the public. These transients are defined as daily visitors and employees who, upon hearing the sirens and receiving the evacuation message, promptly evacuate the EPZ.

Site 2 Cohort 9 represents the schools within the EPZ. Schools receive early and direct notification from OROs allowing them to prepare for evacuation and evacuate earlier than the general public.

Site 2 Cohort 10 represents the transient population within the EPZ. This includes employees who work within the EPZ but do not live within the EPZ, visitors, shoppers, etc. This group is dispersed throughout the EPZ and receives the warning generally at the same time as the public. These transients are defined as daily visitors and employees, who upon hearing the sirens and receiving the evacuation message, promptly evacuate the EPZ.

4.2   WinMACCS Parameters

Standard MACCS2 modeling for NRC assessments uses the parameters in Sample Problem A which is discussed in the MACCS2 User's Manual (NRC, 1998b). For consistency with NRC modeling practices, many of the MACCS2 input parameters used in this study are identical to those in Sample Problem A. Following the DUQI method, a comprehensive list of Sample Problem A parameters was reviewed and appropriate parameters were adjusted as necessary to represent the specific sites being analyzed. Selected parameters that are important to EP are described below with discussion regarding their values for the baseline analysis and the ad hoc analysis.

4.2.1   O-Alarm

O-Alarm is a parameter in the MACCS2 model that defines the time at which notification is given to off-site emergency response officials to initiate protective measures for the EPZ population. For this project, O-Alarm is the time at which OROs sound the sirens. This time is a function of the accident sequence and is measured from the accident initiation.

4.2.2   Evacuation Speeds

As required by 10 CFR 50.47 Appendix E, licensees shall provide an analysis of the time required to evacuate the EPZ. Licensees develop an ETE following the guidance in Appendix 4 to NUREG-0654/FEMA-REP-1, Rev. 1 (NRC, 1980). Additional guidance is provided in NUREG/CR-6863, "Development of Evacuation Time Studies for Nuclear Power Plants," (NRC, 2005b) and NUREG/CR-7002, "Criteria for Development of Evacuation Time Estimate Studies," (NRC, 2011b). ETE studies provide estimated evacuation times for many scenarios during which an evacuation may be implemented. In addition to the estimated evacuation time, these studies contain demographic and evacuation related information regarding the response activities of the general public, transients, and special facilities providing site specific data that can be used in consequence analyses. The following ETEs were used to develop evacuation speeds for the Supplement 3 response.

Table 4-1. Site Specific ETEs

Evacuation	0-2 Mile ETE	2-5 Mile ETE	5-10 Mile ETE**
Site 1: Summer Scenario	ETE_{90} = 0.75 hour ETE_{100} = 1.0 hours	ETE_{90} = 8 hours ETE_{100} = 11 hours	ETE_{90} = 10 hours ETE_{100} = 13 hours
Site 2: Winter Scenario	ETE_{90} = 1 hour ETE_{100} = 1.5 hours	ETE_{90} = 3 hours ETE_{100} = 4 hours	ETE_{90} = 4 hours ETE_{100} = 5 hours

**The ETE for the 5-10 mile area is assumed to be equal to the full ETE for the EPZ.

Currently, site ETEs do not include information regarding staged evacuation, and representative evacuation speeds were therefore calculated based on known information. Future ETEs developed following the guidance in NUREG/CR-7002 are expected to include the time to implement staged evacuations.

It was necessary to develop an ETE for evacuation under ad hoc conditions to develop evacuation speeds. For the ad hoc scenario, it is assumed:

- There are no sirens within the EPZ and notification is conducted via route alerting, EAS messaging, Reverse 911® and other methods. Route alerting is a planned backup for use in the event that sirens are unavailable in areas of an EPZ. It is an effective method for notifying the public and is demonstrated routinely in ad hoc evacuations (NRC, 2005a), but takes longer than sirens to complete notification.

- There is no preplanned traffic control to direct traffic out of the EPZ.

Two approaches were used to determine appropriate ETEs for the ad hoc scenarios. The first was based on information in NUREG/CR-6864 which researched large scale evacuations (NRC, 2005a). Most of the evacuations in that study used route alerting rather than sirens for notification of the public. Route alerting is the primary means of notification in the ad hoc scenario. This resource intensive effort is an effective and proven method successfully used in large scale evacuations. The size of the affected area, number of evacuees, and available resources affect the time to notify the public who then evacuate the area. NUREG/CR-6864 included case studies of large scale evacuations and provided many examples of evacuations conducted using route alerting and supplemental notification techniques (NRC, 2005a). The diversity of incidents studied in NUREG/CR-6864 showed evacuations of as many as 40,000 people in one hour (from a shopping mall and surrounding area) to 45,000 people in 8 hours. Most of the evacuations in the study included populations from about 2,000 to 5,000 people and these occurred in both rural and urban areas. Typically, fewer people per hour were evacuated in rural areas than in urban areas. A direct relationship between time, area, and population density could not be established with the available data; therefore, a scaling factor for evacuation time was estimated. An estimate of 8,000 people per hour was used for urban areas similar to representative Site 1, and an estimate of 5,000 people per hour was used for rural areas similar to representative Site 2. Actual times would be dependent upon the available resources, size of the area, and population density. Using the NUREG/CR-6864 approach and the people per hour values above, evacuation of Site 1 would be estimated to take 18 hours and evacuation of Site 2 would be estimated to take 8 hours.

A second approach to develop ETEs for the ad hoc scenario included review of existing ETEs. Site specific ETEs for many NPP sites include analysis in which times for planned and ad hoc traffic control are both included. A review of selected ETEs showed that an ETE can increase as much as 25 percent when traffic control is unplanned, depending upon the population density and roadway characteristics. An estimate was also developed using the 25 percent increase in time that may be realized if preplanned traffic control is not implemented. Using this approach for the two sites, increasing the ETE by 25 percent provides ETEs of 16.25 hours for Site 1 and 6 hours for Site 2.

The longer evacuation times using number of people per hour (e.g., 18 hours for Site 1 and 8 hours for Site 2), were selected for use in the ad hoc analyses because a larger amount of data was available for consideration. The 90 percent ETEs were calculated using the same ratio as the Supplement 3 response and rounded to the nearest hour. The ETE values are then translated into speeds for each cohort.

The evacuation routes were obtained from the ETE and together with the local grid network were used to establish direction of travel for the evacuees. The travel direction and speed multipliers were input onto the WinMACCS grid. The WinMACCS grid for a generic site is illustrated in Figure 4-3. Using the WinMACCS network evacuation

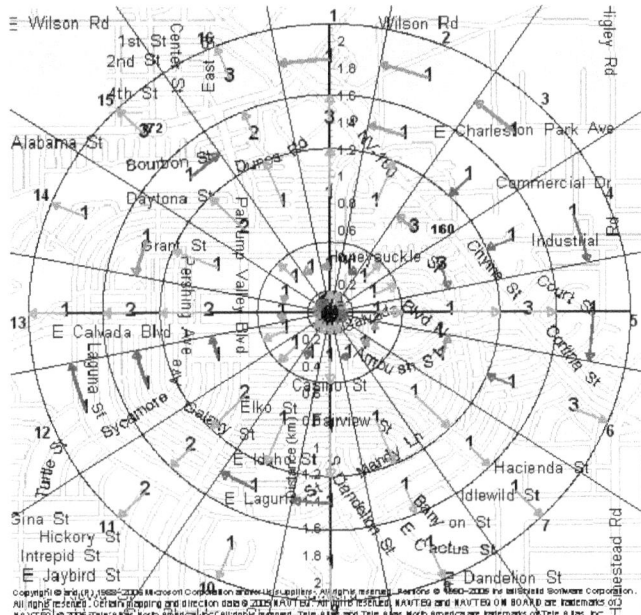

Figure 4-3. WinMACCS grid showing traffic direction arrows and speed multipliers

application, the evacuation was modeled to the EPZ boundary, which was assumed to be 10 miles from the plant. This is a general assumption in the process because WinMACCS uses concentric rings and cannot accept an irregularly shaped EPZ. The user determines the ring distances, which may be greater or less than 10 miles depending on the specific site.

Roadways within an EPZ are not constructed radially away from an NPP, requiring travel of more than 10 miles in some instances. A maximum travel distance of 13 miles was used to develop speeds. This distance was based on review of maps of the EPZ roadway networks which indicate that maximum travel distance would be about 30 percent longer than the radial distance. Consistent with typical MACCS2 analyses, beyond 20 miles the evacuating population received no further dose.

4.2.3 Shielding Factors

Shielding factors vary by geographical region across the United States. The values used for both sites in the analysis are shown in Table 4-2 and represent average values for the region based on previous analyses. The factors represent the fraction of dose that a person would be exposed to when performing normal activities, evacuating, or staying in a shelter in comparison to a person outside with full exposure. Special facilities are typically larger and more robust structures than housing stock and therefore have better shielding factors as indicated. A value of zero indicates complete shielding, and a value of one indicates no shielding.

Table 4-2. Shielding Factors

Cohort	Ground shine			Cloud shine			Inhalation/Skin		
	Normal	Evac.	Shelter	Normal	Evac.	Shelter	Normal	Evac.	Shelter
Normal facilities	0.22	0.50	0.15	0.64	1.00	0.55	0.46	0.98	0.33
Special facilities	0.05	0.50	0.05	0.31	1.00	0.31	0.33	0.98	0.33

The normal activity shielding factors have been adjusted to account for the understanding that people do not spend a great deal of time outdoors. The normal activity values are all weighted averages of indoor and outdoor values based on being indoors 81 percent of the time and outdoors 19 percent of the time (Wheeler, 2000). The shielding factor value for indoor activities was assumed to be the same as the shielding factor value for sheltering.

4.2.4 Potassium Iodide (KI)

The purpose of the KI, as a protective action, is to saturate the thyroid gland with stable iodine so that further uptake of radioactive iodine by the thyroid is diminished. If taken at the right time and in the appropriate dosage, KI can nearly eliminate doses to the thyroid gland from inhaled radioiodine. Factors that contribute to the effectiveness of KI include availability, timing of ingestion, and the degree of pre existing stable iodine saturation of the thyroid gland. The analysis assumes that some residents will not remember where they have placed their KI or may not have it available and will therefore not take KI. It is also assumed some residents will not take their KI when directed (i.e., they may take it early or late which reduces the efficacy). To account for these factors, the analysis assumed that KI is taken by about 50 percent of the public, and the efficacy of the KI was set at 70 percent. For the ad hoc response, it is assumed that no KI is administered.

4.2.5 Hotspot and Normal Relocation

"Hotspot" and "normal" relocation are features of the MACCS2 code that model additional protective actions implemented by OROs. Because this project only investigates the consequences within the EPZ, these relocation criteria are only applied to the non-evacuating cohort. In addition to prompt protective actions, residents would be relocated from areas where the dose exceeds protective action criteria based on EPA Protective Action Guides (PAGs) (EPA, 1992). Some states establish more stringent criteria than the EPA PAGs, but for this project the values were assumed to be the same at each site. OROs would determine the affected areas based on dose projections using State, utility, and Federal agency computer models and field measurements. Hotspot relocation and normal relocation models are included in the MACCS2 code to reflect this activity. These models include dose from cloudshine, groundshine, direct inhalation, and resuspension inhalation. When these models are applied within the MACCS2 calculation, individuals who would be relocated because their projected total committed dose from these pathways exceeds the protective action criteria are prevented from receiving any additional dose during the emergency phase. The relocation dose criterion are applied at a specified time after plume arrival at the affected area.

For this study, hotspot relocation of individuals occurs 12 hours after plume arrival if the total lifetime dose commitment for the weeklong emergency phase is projected to exceed 5 rem (0.05 sievert (Sv)). Normal relocation of individuals occurs 24 hours after plume arrival if the total lifetime dose commitment is projected to exceed 1 rem (0.01 Sv). The dose criteria is based on the upper and lower EPA PAG values. The relocation times of 12 hours for hotspot and 24 hours for normal relocation were estimated considering that OROs may not be available earlier to assist with relocation due to higher priority tasks in the evacuation area. For the ad hoc scenario, normal and hotspot relocation were not applied.

4.2.6 Habitability

Habitability is the consequence model parameter that is used to establish the dose level at which residents are allowed to return to the EPZ to live. Because this study is a comparison of the immediate effects of EP during the early phase, long-term habitability was not used in the analysis.

4.2.7 Adverse Weather

Adverse weather is typically defined as rain, ice, or snow that affects the response of the public during an emergency. The affect of adverse weather on the mobilization of the public was not directly considered in establishing emergency planning parameters for this project. However, adverse weather was addressed in the movement of cohorts within the analysis. The ESPMUL parameter in WinMACCS is used to reduce travel speed when precipitation is occurring as indicated from the meteorological weather file. The ESPMUL factor was set at 0.7, which slows down the evacuating public to 70 percent of the established travel speed when precipitation exists.

4.2.8 Surface Roughness Coefficient

A linear scaling factor is applied to the dispersion formula to adjust the vertical dispersion parameters to account for surface roughness. A single coefficient is used in the modeling. The surface roughness coefficient selected was 60 cm for each site to represent woodland forest type areas intermixed with suburban areas. A value of 10 cm represents grassland, whereas a value of 100 cm is representative of the forest areas and urban type areas.

5.0   CONSEQUENCE ANALYSES

5.1   Response Scenarios

The accident sequences and resulting response scenarios were developed specifically for the two reference sites. The suggested protective action recommendation paths in the following descriptions are based on the hypothesized accident scenarios. Site specific PARs would be expected to consider onsite and offsite information that may influence the decision path, which may differ from the paths analyzed below. This demonstration project identifies the bases for the selected paths and completes the analyses under the established conditions.

This analysis is for hypothetical sites. Actual site emergency plans would be used in any regulatory regimen and may vary from the assumptions used here, e.g., in this study it is assumed that SIP would be used for hostile action based events. A site specific plan might not use that logic. Alternately, if SIP is used for hostile action, the OROs might still evacuate populations separated from the plant by natural barriers such as rivers. These variations are not addressed as this study is intended to provide a proof of concept suitable for further consideration.

5.2   Supplement 3 and Ad Hoc Response

The Supplement 3 response represents the baseline analysis and was developed assuming that the response to the postulated accident is consistent with the activities identified in the onsite and offsite emergency response plans, which would use the Protective Action Logic Diagram in Supplement 3 to NUREG-0654/FEMA-REP-1, Rev. 1 shown in Figure 5-1 (NRC, 2011a). It is assumed that the onsite and offsite emergency response plans are implemented, and the public responds to protective actions when they are received. Timing is an important factor in EP, therefore, discussions are developed around the timeline of events. A timeline is developed for each accident sequence to represent onsite and offsite decisions and the expected response of the public. The analysis takes credit for the physical and administrative notification capabilities that are established. Offsite emergency plans include provisions for evacuating the general public, schools, transients (e.g., visitors), and special facilities from the EPZ. Traffic control would be established to facilitate the evacuation.

The ad hoc evacuation is intended to postulate a response that might occur if there were no onsite or offsite emergency plans specifically developed for an emergency at the NPP. The OROs would be expected to respond similarly to a response to any other emergency in the area. For example, when OROs initiate an evacuation in response to a chemical plume, they typically evacuate downwind and expand the evacuation if needed (NRC, 2005a). This requires broadcasting an EAS message and implementing route alerting and other available notification methods (e.g., Reverse 911® type). The evacuation is assumed to start within 15 minutes of the broadcast of the EAS message and will increase as route alerting expands throughout the EPZ. For the ad hoc analysis, it is assumed that OROs initially evacuate to 5 miles and then expand the evacuation to 10 miles. This was simulated in the model by evacuating the population of the outer rings of the EPZ at a slower rate.

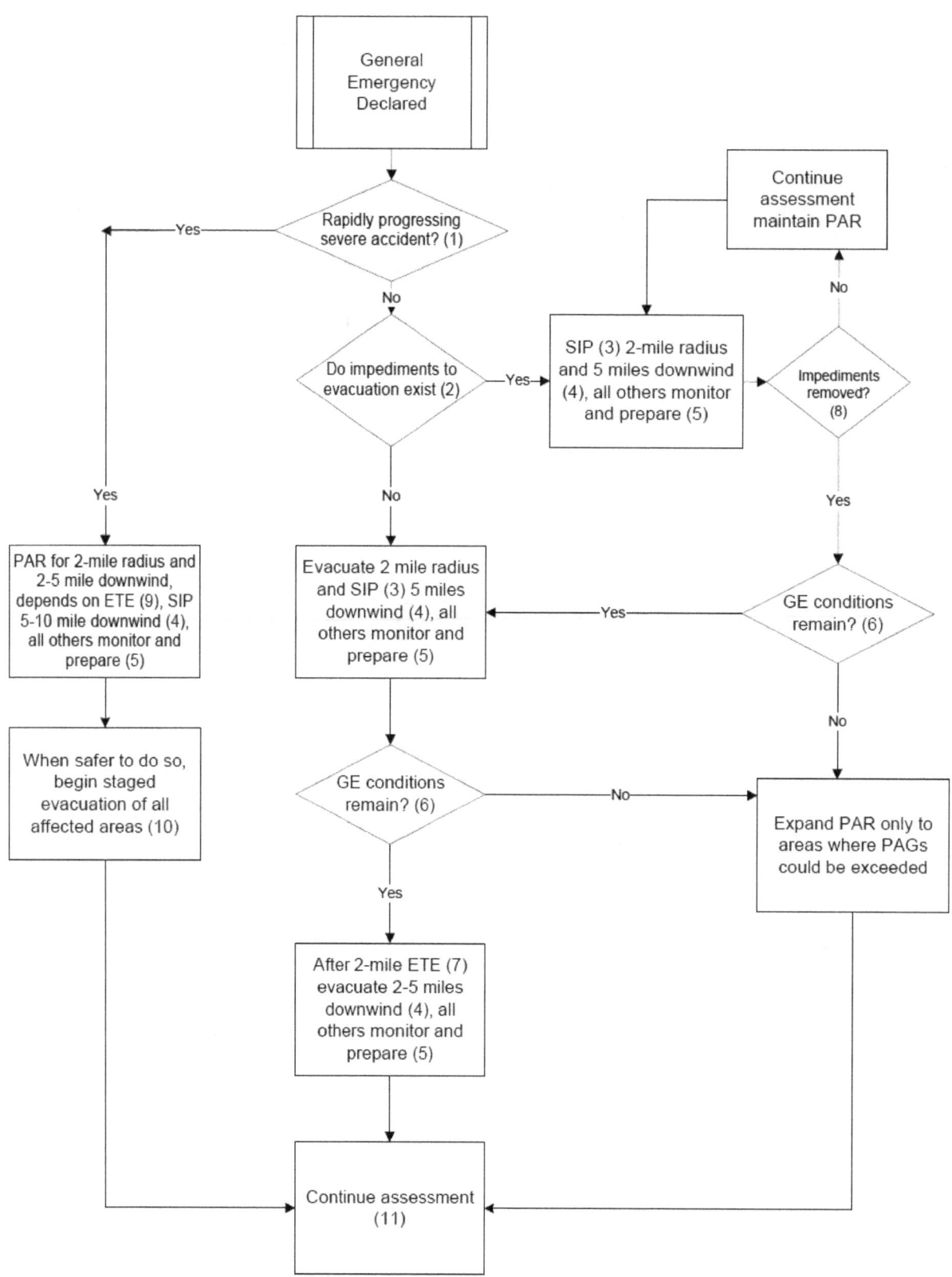

Figure 5-1. Protective action strategy development tool**
**The numeric notes in the chart may be found in Supplement 3
(NRC, 2011a)

5.2.1 Reference Site 1 Supplement 3 Response - STSBO

The PAR strategy for the Site 1 STSBO is not a rapidly progressing severe accident and should follow the center column of the Protective Action Strategy Development Tool in Figure 5-1 (NRC, 2011a). The STSBO PAR strategy for this analysis is:

- Evacuate the 0-2 mile area and SIP 5 miles downwind. Monitor and prepare in all other areas of the EPZ;
- At 45 minutes, which is the 90 percent 0-2 mile ETE, evacuate 2-5 miles downwind, if necessary. Monitor and prepare 5-10 miles downwind; and,
- Continue assessment and if necessary, evacuate 5-10 miles downwind when appropriate.

OROs initiate the process to notify and evacuate the public after receipt of the General Emergency (GE) declaration. This notification action, which includes sounding sirens and broadcast of the EAS message, is estimated to take approximately 45 minutes which is consistent with exercise data. For this analysis, it is assumed evacuation of the 5-10 mile area begins 2 hours after the start of the 2-5 mile area evacuation. Table 5-1 identifies accident specific response timeline activities. Figure 5-2 displays the timeline of response activities for the accident scenario providing a representation of cohort movements.

Table 5-1. Site 1 Supplement 3 Response STSBO-TI SGTR

Time	Activity
0:00	Initiating Event
0:15	Plant declares a site area emergency (SAE) and notifies OROs. OROs initiate offsite notifications to support agencies and special facilities.
2:00	Plant declares GE and notifies OROs.
2:45 O-Alarm	OROs sound sirens and broadcast EAS message. Initial PAR is evacuation of 2 miles and SIP 2-5 miles downwind. Monitor and prepare all other areas of the EPZ. Transient 1 and Transient 2 evacuate immediately. These facilities would have been notified directly by OROs after SAE was declared. The 0-2 mile general public begins to evacuate.
3:30	ETE90 for the 0-2 mile area is 45 minutes for this site, at which time the 2-5 mile downwind general public is instructed to evacuate. SIP is instructed for the 5 to 10 mile area downwind. Transient 3 begins to evacuate.
5:00	Shadow evacuation begins. By this time residents in the shadow area have observed large numbers of EPZ residents evacuating and have followed media reports covering the emergency. The shadow is modeled as evacuating at a specific time. An actual shadow evacuation would be spread over a period of time.
5:30	After about 2 hours, begin evacuating the 5-10 mile area. At this time, evacuation of the 2-5 mile area is well underway.
7:00	Special facilities evacuate. Special facilities are modeled as evacuating at a specific time. An actual evacuation of special facilities would occur over a period of time based on mobilization needs and availability of transportation resources.
9:00	Special needs residents evacuate. Special needs residents are modeled as evacuating at a specific time. An actual evacuation of special needs residents would occur over a period of time based on mobilization needs and availability of transportation resources.

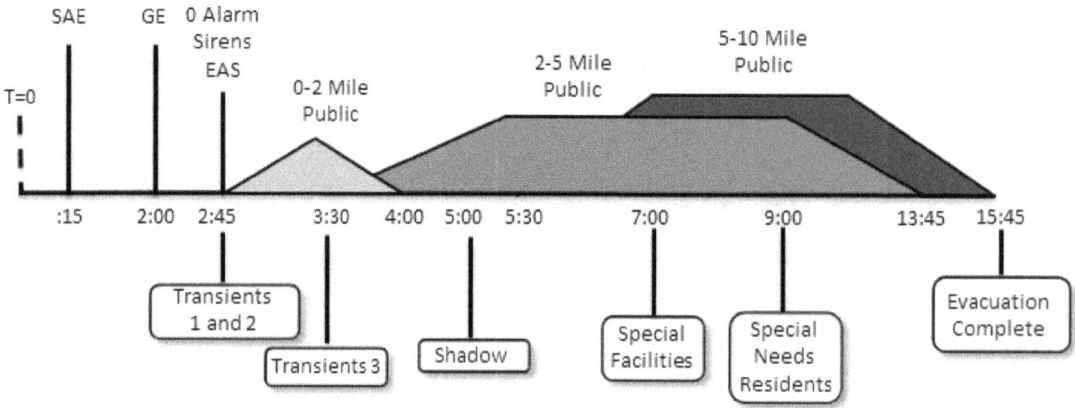

Figure 5-2. Site 1 Supplement 3 Response Timeline: STSBO

5.2.2 Reference Site 1 Supplement 3 Response - LBLOCA

The LBLOCA scenario for this project is not a rapidly progressing severe accident and should follow the center column of the Protective Action Strategy Development Tool in Figure 5-1. For this scenario it is assumed impediments to evacuation exist. The PAR strategy from Figure 5-1 would answer "Yes" at the decision box: "Do impediments to evacuation exist," and shelter in place for the 2 mile area and 5 mile keyhole would be the suggested PAR strategy, with all other areas of the EPZ asked to monitor and prepare. This strategy was modeled as though implemented until field dose measurements and dose projections indicate that the plume has passed and evacuation may begin. With the impediments removed and the GE conditions still remaining, the PAR strategy would then continue back down the center column of the Protective Action Strategy Development Tool as follows:

- Evacuate the 0-2 mile area and SIP 5 miles downwind. Monitor and prepare all other areas of the EPZ;
- At 45 minutes, which is 90 percent 0-2 mile ETE, evacuate 2-5 miles downwind, provided it is safe to do so. Monitor and prepare all other areas of the EPZ; and,
- Continue assessment and if necessary, evacuate 5-10 miles downwind when appropriate.

This proof of concept analysis is for a hypothetical site and assumes that OROs have structured protective resonse strategies to SIP when a hostile action event occurs. However, that stucture is up to OROs and may not be used at all sites. Further, the protective action logic implemented by OROs may include evacuating the public more distant from the site, or from areas that are separated by natural barriers, such as rivers. Areas such as these are not likely affected by the hostile action, nor would these evacuations impede the response. These strategies would likely improve results, but were not considered as this analysis is not site specific.

Table 5-2 identifies accident specific response timeline activities. Figure 5-3 illustrates the timeline of response activities for the accident scenario.

Table 5-2. Site 1 Supplement 3 Response LBLOCA

Time	Activity
0:00	Potential issue confirmed by security. Control room notified. Early onsite actions initiated. Reactor scram.
0:15	Plant declares SAE and notifies OROs. OROs initiate offsite notifications (e.g., police, special facilities, etc.).
0:30	Plant declares GE, again notifies OROs
1:00	Initiating event.
1:15 O-Alarm	OROs sound sirens and broadcast EAS message. Initial PAR is to SIP the entire EPZ.
3:15	Transients 2 and 3 begin evacuating.
3:45	Transient 1 begins evacuating.
4:15	Shadow evacuation begins.
6:15	Assume a 5 hour SIP would be implemented at this time. Field dose and modeling confirm when SIP may be ended. Initiate a staged evacuation. Instructions for the 0-2 mile area are to evacuate and all others SIP. The 2-5 mile downwind area prepares to evacuate. SIP is assumed to be directed for the 5 to 10 mile area downwind. Special facilities residents begin to evacuate.
7:00	ETE90 for the 0-2 mile area is 45 minutes, at which time the 2-5 mile downwind general public is instructed to evacuate. SIP is instructed for the 5-10 mile area downwind.
8:15	Special needs residents start to evacuate.
9:00	After about 2 hours, when the 2-5 area evacuation is established and well under way, evacuation of the 5-10 mile area begins.

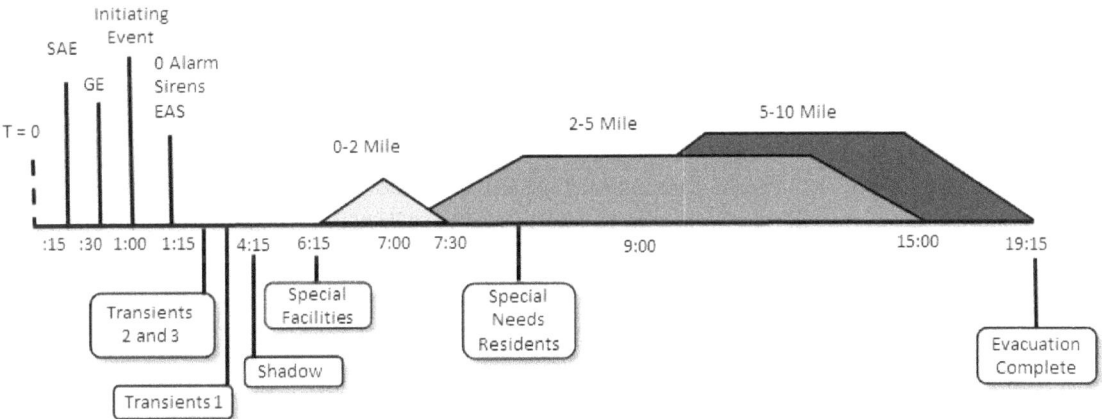

Figure 5-3. Site 1 Supplement 3 Response Timeline: LBLOCA

5.2.3 Reference Site 1 Ad Hoc Response - STSBO

In the ad hoc response scenario, the initial assumption is that there is no formal radiological emergency preparedness program. Given the long history of nuclear plant emergency planning and involvement of OROs, such an assumption may be difficult to envision, but is necessary to establish a baseline ad hoc analysis. The study assumes there are no direct communication links with OROs nor practiced notification protocols. It is expected some form of offsite notification would be required of any industrial facility and plant workers would contact family members. The accident would become known through the societal communications, but activation of ORO organizations, the full briefing of decision makers, and the decision to evacuate would be delayed significantly.

For the ad hoc scenario, response times were adjusted to reflect delays in notification. The shadow population was delayed to represent a slower communication of the emergency. The general public response was modeled as starting slowly and having a longer duration for the evacuation tail. It is assumed that ORO decision makers, responsible for evacuating the public, become aware of the event at about 4 hours and issue evacuation direction about 90 minutes later.

The population fractions for each general public cohort were maintained the same as the baseline. The evacuation of special facilities was adjusted to start 2 hours later than the baseline to account for the additional time for the facilities to become aware of the emergency. The ad hoc response timelines for the Site 1 STSBO and LBLOCA are described in Table 5-3 and Table 5-4 and illustrated in Figure 5-4 and Figure 5-5.

Table 5-3. Site 1 Ad hoc Response STSBO

Time	Activity
0:00	Initiating Event
0:15	Plant onsite response underway.
4:00	ORO decision makers become aware.
4:30	Assume OROs contact plant to confirm there is an emergency.
5:30	Protective action decision is to evacuate based on discussions with the plant regarding the potential for release of radioactive material.
6:15 O-Alarm	EAS message is broadcast and route alerting begins. Alerting expands throughout the 10 mile EPZ.
6:45	General public 1 begins to evacuate. With little understanding of the severity of the emergency and no previous awareness, this cohort is modeled as evacuating 30 minutes after the alerting begins. The cohort is modeled as evacuating at a specific time. In an actual evacuation, this group of the public would evacuate over a period of time.
7:15	General public 2 begins to evacuate.
8:45	General public 3 begins to evacuate. As the distance from the plant increases, the area increases proportional to the square of the distance. The time to complete the route alerting in the progressively larger area takes longer.
9:00	Transient 2 begins to evacuate.
10:00	Transient 1 begins to evacuate.
10:15	Shadow evacuation begins to evacuate.
12:00	Transient 3 begins to evacuate.
13:15	General public 4 begins to evacuate. Special facilities begin to evacuate.
15:15	Special needs residents begin to evacuate.
16:15	The tail of the evacuation begins to evacuate.

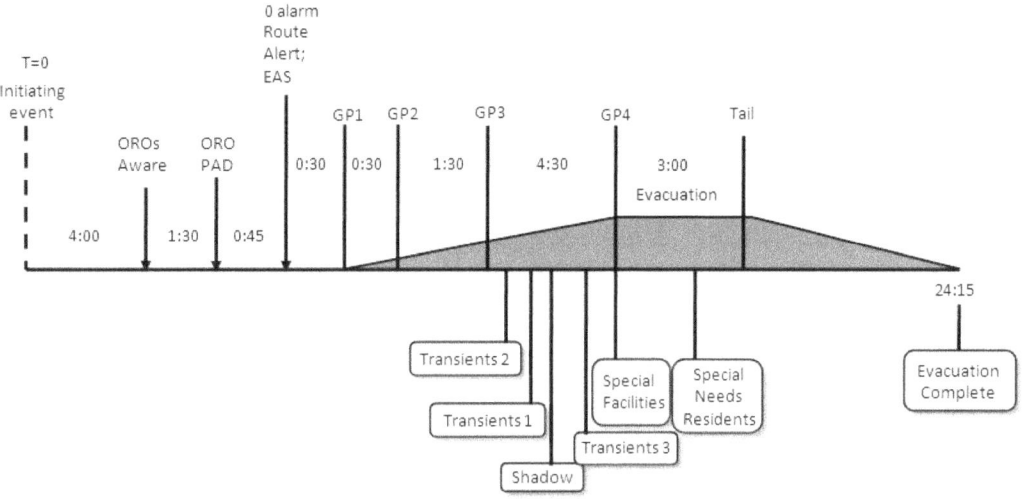

Figure 5-4. Site 1 Ad hoc Response Timeline: STSBO

5.2.4 Reference Site 1 Ad Hoc Response - LBLOCA

The LBLOCA scenario represents an accident in which local law enforcement would be requested to support onsite security. In the ad hoc scenario, it is assumed the site would request police assistance, thus emergency management OROs would be aware of the event promptly, but it is assumed that without preplanning, the protective action decisions will take time to develop. Media attention will begin to alert the public, and response agencies will ultimately issue protective action orders. Evacuation is modeled as taking longer than the Supplement 3 response because the ad hoc scenario assumes the protective action decisions take time to develop, there are no sirens for prompt notification, preplanned traffic control, or prescribed EAS messaging to direct the evacuation.

Table 5-4. Site 1 Ad hoc Response LBLOCA

Time	Activity
0:00	Security incident onsite. No early emergency response actions.
0:15	Plant requests offsite assistance from police.
1:00	Initiating event.
1:30	OROs are aware through police involvement that an emergency at the plant exists.
3:30	ORO decision makers confirm an emergency exists that may threaten the public. Protective action decision based on discussions with the plant regarding potential for release of radioactive material. Decision to evacuate downwind to a distance of 5 miles.
4:15 O-Alarm	EAS message is broadcast and route alerting begins.
4:45	General public 1 begins to evacuate.
5:15	General public 2 begins to evacuate.
6:45	General public 3 begins to evacuate.
7:00	Transient 2 begins to evacuate.
8:00	Transient 1 begins to evacuate.
8:15	Shadow evacuation begins.
10:00	Transient 3 begins to evacuate.
11:15	General public 4 begins to evacuate. Special facilities begin to evacuate.
13:15	Special needs residents begin to evacuate.
14:15	The tail of the evacuation begins to evacuate.

47

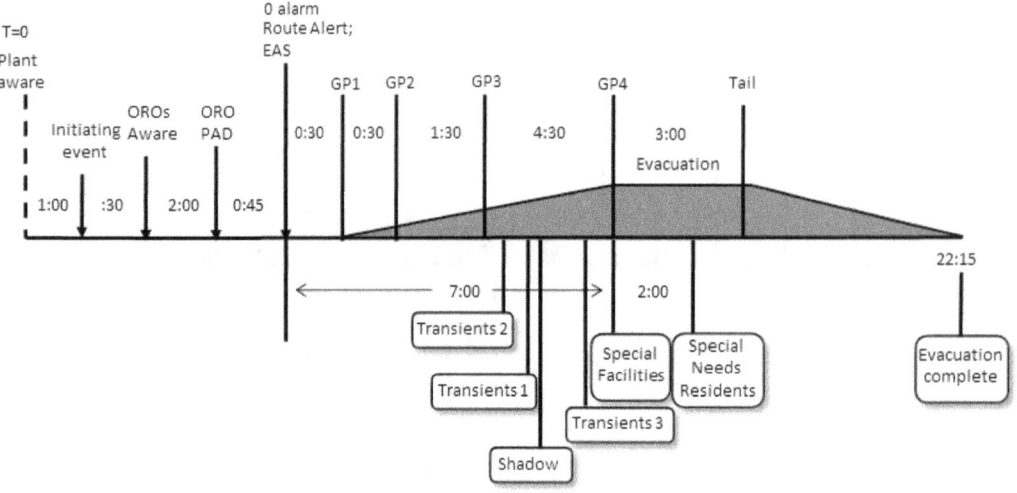

Figure 5-5. Site 1 Ad hoc Response Timeline: LBLOCA

5.2.5   Reference Site 2 Supplement 3 Response - STSBO

The STSBO scenario for this project is not a rapidly progressing severe accident and should follow the center column of the Protective Action Strategy Development Tool in Figure 5-1. The PAR strategy for the Site 2 STSBO includes:

- Evacuate the 0-2 mile area and SIP 5 miles downwind. Monitor and prepare all other areas;
- At 1 hour, which is the 90 percent 0-2 mile ETE, evacuate 2-5 miles downwind. Monitor and prepare all other areas of the EPZ; and,
- Continue assessment and if necessary, evacuate 5-10 miles downwind when appropriate.

There is no appreciable population within the 0-2 mile zone for this site; however, there is a large transient population that would be directed via EAS messaging to evacuate with the initial protective action. At Site 2, OROs mobilize school buses after receipt of the SAE emergency declaration in order to promptly evacuate schools if the accident escalated to a GE. However, there is no SAE in this scenario. School buses would be summoned shortly after notification of the GE and are assumed to be evacuating students within 45 minutes. Upon declaration of a GE, the sirens would be sounded and an EAS message would be broadcast that would include protective action instructions. It is estimated that the sirens and EAS messaging occur approximately 45 minutes after the GE is declared. Table 5-5 identifies accident specific response timeline activities. Figure 5-6 illustrates the timeline of response activities for the accident scenario.

Table 5-5. Site 2 Supplement 3 Response STSBO

Time	Activity
0:00	Initiating Event
0:15	Plant declares immediate GE and notifies OROs.
1:00 O-Alarm	OROs sound sirens and broadcast EAS message. Protective action decision to evacuate 0-2 mile area and SIP 2-5 downwind. Evacuation of general public begins. Monitor and prepare all other areas. Schools evacuate.
2:00	2-5 mile downwind general public starts to evacuate at the ETE90 for the 0-2 mile area which is 1 hour after EAS message for this site. SIP is assumed to be directed for the 5 to 10 mile area downwind. Transients evacuate.
3:30	Shadow evacuation
4:00	Assumed that after about 2 hours, begin evacuating the 5-10 mile area. At this time, evacuation of the 2-5 mile area is well underway. Special facilities evacuate.
4:30	Special needs residents begin to evacuate.

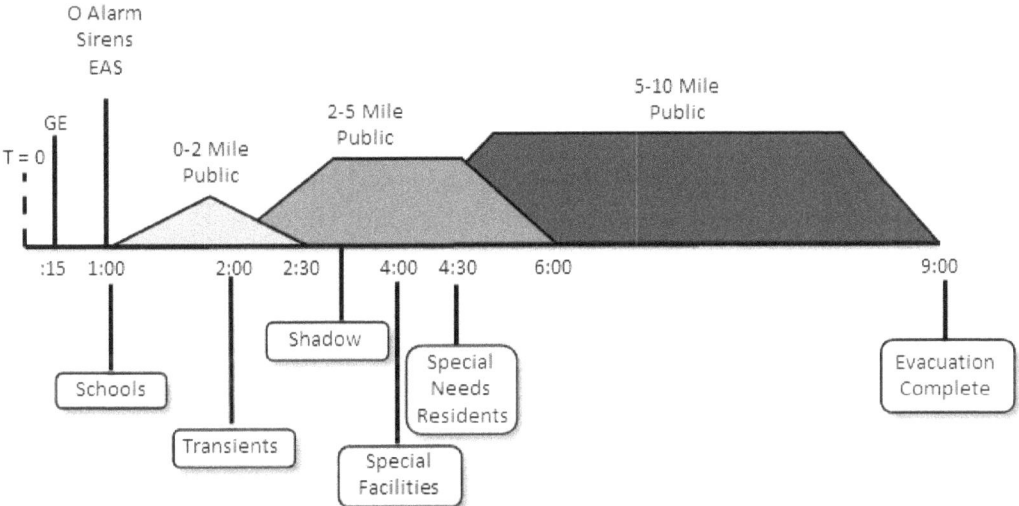

Figure 5-6. Site 2 Supplement 3 Response Timeline: STSBO

5.2.6 Reference Site 2 Supplement 3 Response - ISLOCA

Site 2 ISLOCA is a self revealing rapidly progressing severe incident with rapid loss of containment integrity and loss of all ability to cool the core. For a rapidly progressing severe accident, defined as an immediate GE with rapid loss of containment integrity and loss of all ability to cool the core, the PAR strategy identified in the Protective Action Strategy Development Tool in Figure 5-1 is the left column and includes:

- Evacuate 0-2 mile area because the 90 percent ETE for the 0-2 mile area is 1 hour, and evacuate the 2-5 mile downwind area because the 90 percent ETE for this area is 3 hours. The initial protective action is for the entire keyhole area because the ETE90 for this area is 3 hours. SIP for 5-10 mile downwind sectors.

The site was modeled with an SIP of the 5-10 mile area for an additional 2 hours followed by evacuation. Table 5-6 identifies accident specific response timeline activities. Figure 5-7 illustrates the timeline of response activities for the accident scenario.

Table 5-6. Site 2 Supplement 3 Response ISLOCA

Time	Activity
0:00	Potential issue confirmed by security. Control room notified. Early onsite actions initiated. Reactor scram.
0:05	Plant provides OROs immediate notification and maintains communication.
0:15	Initiating event – self revealing. Plant declares GE and notifies OROs.
1:00 O-Alarm	OROs sound sirens and broadcast EAS message. Protective action decision is to evacuate the 0-2 mile area and the evacuation of the downwind 2-5 mile keyhole. SIP 5-10 mile area, all others monitor and prepare. The ETE90 for the 0-5 mile area is 3 hours, therefore following the PAR Logic Diagram, the 2-5 mile downwind area is also directed to evacuate with the initial protective action decision. Schools begin to evacuate.
2:00	Transients evacuate.
2:30	Shadow evacuation.
3:00	After about 2 hours evacuation of the 5-10 mile area begins.
4:00	Special facilities evacuate.
4:30	Special needs residents evacuate.

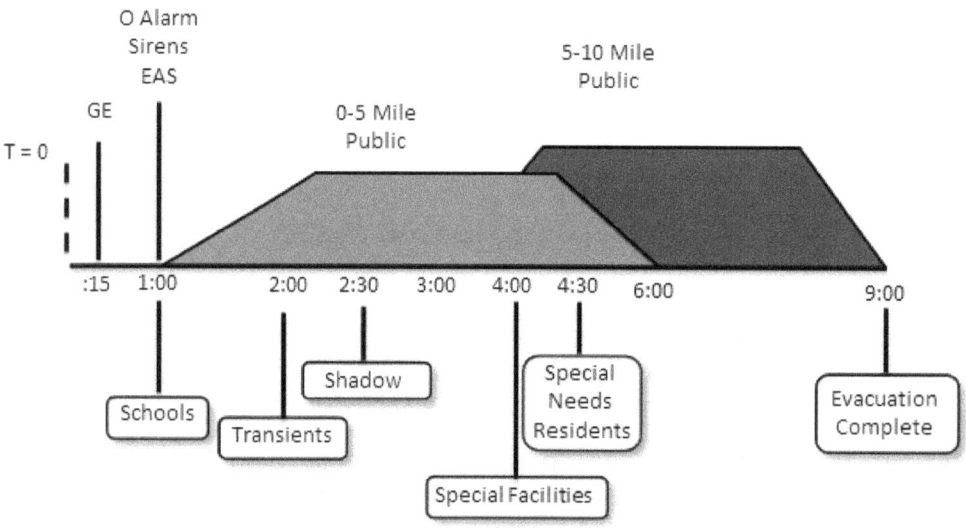

Figure 5-7. Site 2 Supplement 3 Response Timeline: ISLOCA

5.2.7   Reference Site 2 Ad Hoc Response - STSBO

For the ad hoc scenario, response times were adjusted to reflect delays in notification. The shadow population was delayed to represent a slower communication of the emergency. The general public response was modeled starting slowly and having a longer duration for the evacuation tail. The population fractions for each general public cohort were maintained the same as the baseline. The evacuation of special facilities was adjusted to start 2 hours later than the baseline to account for the additional time for the facilities to become aware of the emergency. The ad hoc response timelines for the Site 2 are described in Table 5-7 and Table 5-8 and illustrated in Figure 5-8 and Figure 5-9.

Table 5-7. Site 2 Ad hoc Response STSBO

Time	Activity
0:00	Initiating Event
0:15	Plant onsite response underway.
4:00	OROs become aware through societal communication as workers contact family members.
4:30	OROs contact plant to confirm there is an emergency.
5:30	ORO decision makers issue a protective action decision based on discussions with the plant regarding the potential for release of radioactive material. Decision is to evacuate.
6:15 O-Alarm	EAS message is broadcast and route alerting begins.
6:45	General public 1 begins to evacuate.
7:15	General public 2 begins to evacuate.
8:00	Schools begin to evacuate.
8:15	Shadow evacuation begins.
8:45	General public 3 begins to evacuate.
9:15	General public 4 begins to evacuate. Transients begin to evacuate.
9:45	The tail of the evacuation begins to evacuate.
11:45	Special facilities begin to evacuate.
13:15	Special needs residents begin to evacuate.

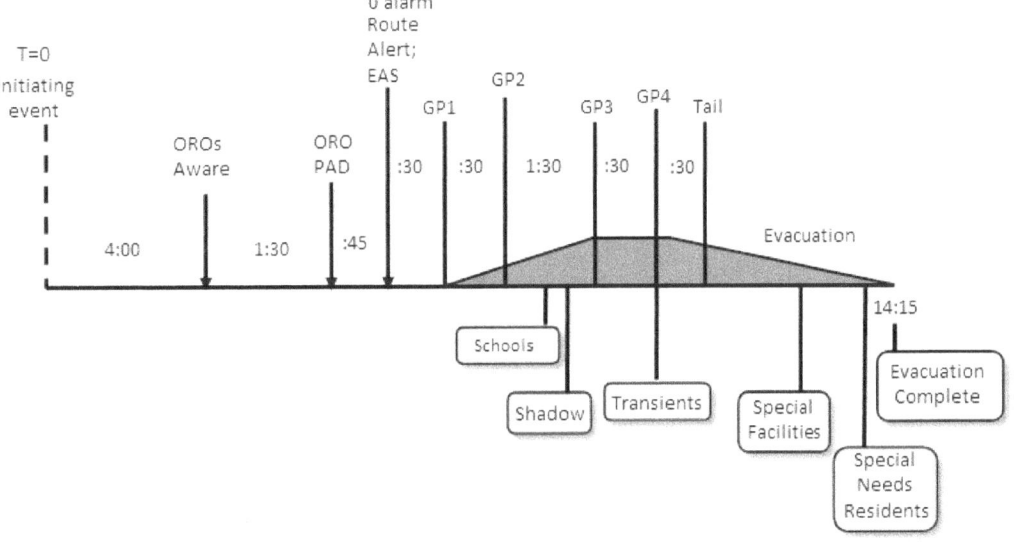

Figure 5-8. Site 2 Ad hoc Response Timeline: STSBO

5.2.8   Reference Site 2 Ad Hoc Response - ISLOCA

The ISLOCA is a self revealing accident onsite involving a fire of such scale that fire department assistance would be requested promptly, making local OROs aware of the event. In the ad hoc scenario, it is assumed that although OROs would be aware of the event promptly, without preplanning, the protective action decisions will take time to develop. Media attention will begin to alert the public, and response agencies will ultimately issue protective action orders. Evacuation is modeled as taking longer than the Supplement 3 response because the ad hoc scenario assumes the protective action decisions take time to develop, there are no sirens for prompt notification, preplanned traffic control, or prescribed EAS messaging to direct the evacuation.

Table 5-8. Site 2 Ad hoc Response ISLOCA

Time	Activity
0:00	Initiating Event – self revealing.
0:15	Plant onsite response underway. OROs are aware of incident.
0:30	OROs contact the plant and confirm emergency.
1:30	Protective action decision based on discussions with the plant regarding the potential for release of radioactive material.
4:15 O-Alarm	EAS message is broadcast and route alerting begins.
4:45	General public 1 begins to evacuate.
5:15	General public 2 begins to evacuate.
6:00	Schools begin to evacuate.
6:15	Shadow evacuation begins.
6:45	General public 3 begins to evacuate.
7:15	General public 4 begins to evacuate. Transients begin to evacuate.
7:45	The tail of the evacuation begins to evacuate.
9:45	Special facilities begin to evacuate.
11:15	Special needs residents begin to evacuate.

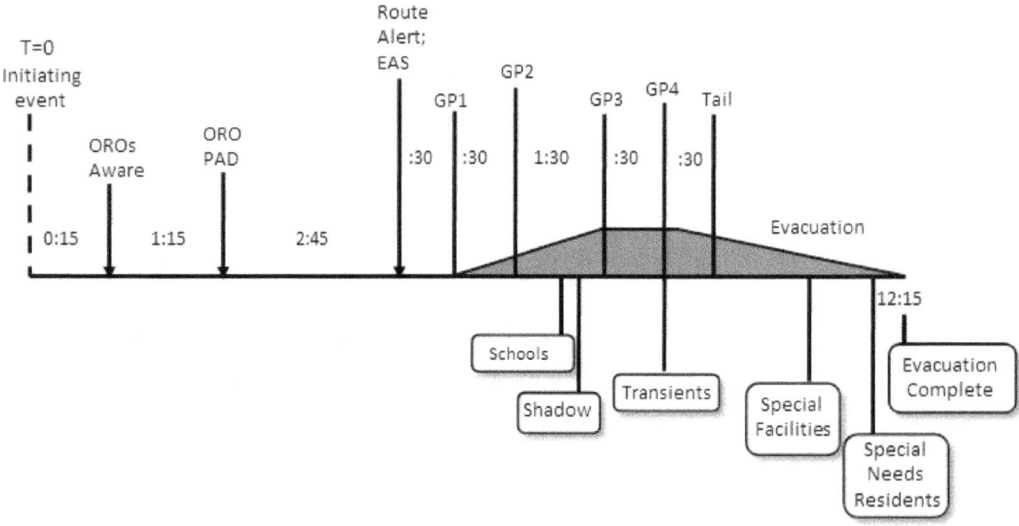

Figure 5-9. Site 2 Ad hoc Response Timeline: ISLOCA

5.3 Value of Emergency Preparedness

After completion of the Supplement 3 and ad hoc analyses for each of the sites and accident sequences, a comparison between the baseline and ad hoc results was performed. The metric for comparison was cumulative population dose within the EPZ. The dose was measured for the Early Phase only, which was set as 7 days. The results are presented in Table 5-9 and

52

illustrated in Figure 5-10. These results show the cumulative dose is greater for the ad hoc response than the Supplement 3 response for every scenario. The Site 1 STSBO shows an increase of about one order of magnitude in dose between the Supplement 3 response and an ad hoc response. For the Site 1 LBLOCA, the increase in dose is small. The Site 2 STSBO shows an increase of about 2 orders of magnitude in dose between the Supplement 3 response and an ad hoc response. For the Site 2 ISLOCA the increase in dose is about 20 percent.

These results illustrate the value of EP in terms of dose avoided by the public through implementation of an EP program and shown that EP may be amenable to being risk-informed. This comparison provides perspective on the magnitude of the risk impact of the EP regulatory framework.

Table 5-9. Cumulative Population Dose for Supplement 3 and Ad Hoc Response

Sequence	Supplement 3	Ad Hoc
Site 1 STSBO	1.78×10^5	3.67×10^5
Site 1 LBLOCA	3.37×10^6	3.62×10^6
Site 2 STSBO	1.65×10^3	1.97×10^5
Site 2 ISLOCA	2.64×10^6	3.20×10^6

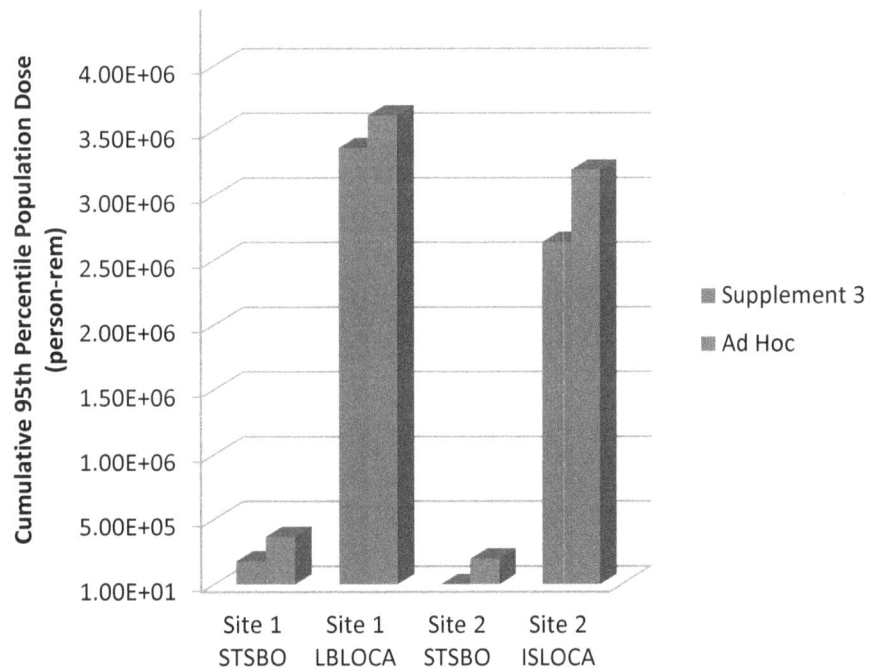

Figure 5-10. Cumulative Population Dose for Supplement 3 and Ad Hoc Response

6.0   DEDUCTIVE QUANTIFICATION INDEX (DUQI)

In the previous section, a method was used to quantify the value of EP in terms of avoided population dose. The next step was to determine whether the DUQI method could be used to quantify inidividual EP program elements. Using the STSBO accident sequences, the following EP elements were evaluated:

- **Siren Scenario**: Response is modeled considering that the EPZ siren system is not operable in the 2-5 mile area around the plant.
- **Notification Delay Scenario**: Response is modeled with a delay of 1 hour in the implementation of protective actions.

The main components of the DUQI method are identified in Figure 6-1.

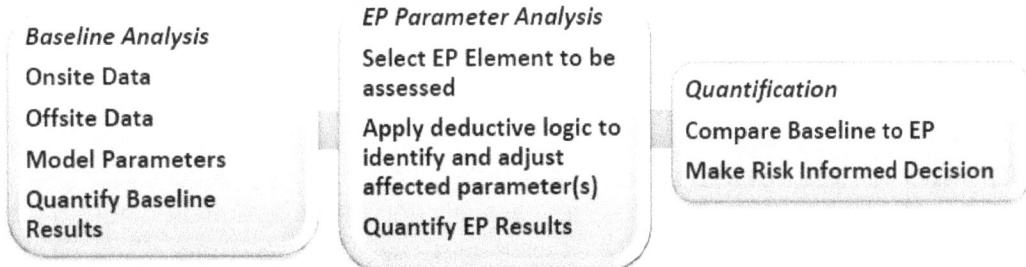

Baseline Analysis

Onsite Data

Offsite Data

Model Parameters

Quantify Baseline Results

EP Parameter Analysis

Select EP Element to be assessed

Apply deductive logic to identify and adjust affected parameter(s)

Quantify EP Results

Quantification

Compare Baseline to EP

Make Risk Informed Decision

Figure 6-1. Main components of the DUQI method

6.1   Siren Scenario

In this scenario, the Supplement 3 PAR is modeled, and it is assumed that sirens in the 2-5 mile area fail unexpectedly. All other EP elements are unchanged from the Supplement 3 response model. The assumed lack of sirens in the 2-5 mile area may be expected to cause the evacuation of this area to take longer because initial alert and notification of the public in this area is delayed. Any change in consequences when compared to the baseline Supplement 3 response should represent the value of the sirens in the 2-5 mile area.

The scenario was developed assuming that the response to the postulated accident is consistent with the activities identified in the onsite and offsite emergency response plans, except for the siren failure. In this scenario, the plant makes a formal declaration of the event and promptly notifies OROs consistent with the Supplement 3 response timelines. Offsite emergency plans are implemented and include provisions for evacuating residents, schools, special facilities and others from the EPZ. EAS messaging is broadcast throughout the EPZ, including the 2-5 mile area. Traffic control is established to facilitate evacuation out of the EPZ. It is assumed the public within the 2-5 mile area receive the alert and notification via route alerting, EAS messages, and societal communication.

Following the DUQI method, the analyst identifies the affected response parameters and determines the appropriate parameter values for the scenario. For the Supplement 3 response, sirens are sounded throughout the EPZ and an ETE was available to support development of travel speeds. In the area from 0-2 miles, the response is identical to the Supplement 3 response. As residents in the 2-5 mile area become aware of the need to evacuate, they will

55

load the roadway network over a longer period of time which will initially result in less congestion and slightly faster speeds, but a longer overall evacuation time.

The parameters affected in this scenario include:

- Delay to shelter (DLTSHL) which is the delay from the time of the start of the accident until the public enters the shelter.
- Delay to evacuation (DLTEVA) which is the length of the sheltering period from the time the public enters the shelter until the point at which they begin to evacuate.
- The evacuation speed (ESPEED) which is assigned for each of the three phases used in WinMACCS including Early, Middle, and Late.

To calculate the appropriate values for the above parameters, an ETE for this scenario was developed using information from the Supplement 3 and ad hoc scenarios. The ETE developed for the ad hoc scenario considered a condition in which there are no sirens and route alerting was the method of alert and notification throughout the 10 mile EPZ. The 2-5 mile area represents about 21 percent of the EPZ area. It is expected that the ETE for siren scenario would be longer than the Supplement 3 response and shorter than the ad hoc response. The ETEs for the no siren scenario were set at 75 percent of the ad hoc scenario and are presented in Table 6-1. The speeds were developed from these ETEs. The response timelines for sites 1 and 2 are described in Table 6-2 and Table 6-3 respectively and are illustrated in Figure 6-2 and Figure 6-3 respectively.

Table 6-1. Scenario ETEs: No Sirens 2-5 Miles.

Evacuation	100 percent ETE	90 percent ETE
Site 1	16 hours	13 hours
Site 2	6.25 hours	5 hours

Table 6-2. Reference Site 1 STSBO No Siren Scenario

Time	Activity
0:00	Initiating Event
0:15	Plant declares SAE and notifies OROs. OROs initiate offsite notifications to support agencies, special facilities.
2:00	Plant declares GE and notifies OROs.
2:45 O-Alarm	OROs sound sirens and broadcast EAS message. Initial PAR is evacuation of 2 miles and SIP 2-5 miles downwind. Monitor and prepare all other areas of the EPZ. Transient 1 and Transient 2 evacuate immediately. These facilities would have been notified directly by OROs after SAE was declared. The 0-2 mile area general public begin to evacuate.
5:00	ETE90 for the 0-2 mile area is 45 minutes for this site, at which time the 2-5 mile downwind general public is instructed to evacuate via route alerting. This is delayed due to lack of sirens. SIP is instructed for the 5 to 10 mile area downwind.
5:30	At this time, evacuation of the 2-5 mile area is well underway and evacuation of the 5-10 mile area begins. Transient 3 evacuates. Shadow evacuates at this time.
7:00	Special facilities, notified via telephone, have mobilized resources and evacuate at this time. Special facilities are modeled as evacuating at a specific time. An actual evacuation of special facilities would occur over a period of time based on mobilization needs and availability of transportation resources.
9:00	Special needs residents, notified via telephone, have mobilized resources and evacuate at this time. Special needs residents are modeled as evacuating at a specific time. An actual evacuation of special needs residents would occur over a period of time based on mobilization needs and availability of transportation resources.

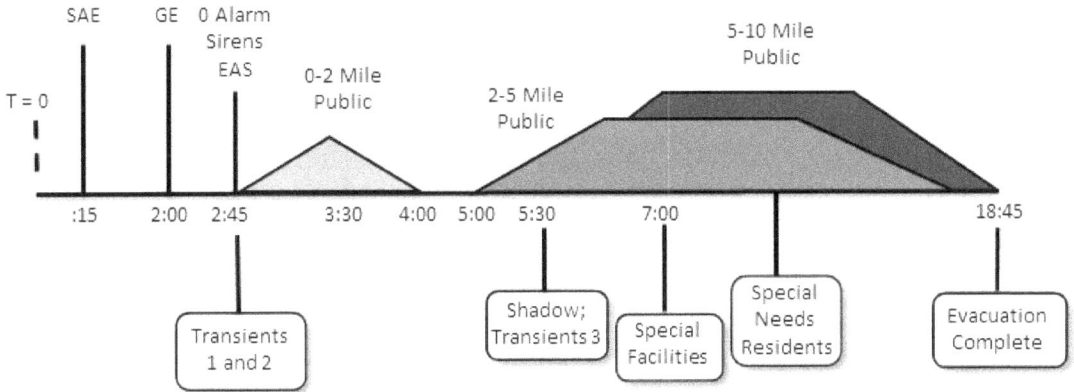

Figure 6-2. Site 1 No Siren Scenario Response Timeline

Table 6-3. Reference Site 2 STSBO No Siren Scenario

Time	Activity
0:00	Initiating Event
0:15	Plant declares immediate GE and notifies OROs.
1:00 O-Alarm	OROs sound sirens and broadcast EAS message. In this scenario, OROs make a protective action decision to evacuate 0-2 mile area and SIP 2-5 downwind. Schools evacuate. Special facilities and special needs are notified to prepare. Monitor and prepare all other areas.
3:15	2-5 mile downwind general public begins to evacuate having been notified via route alerting and EAS messaging.
4:00	It is assumed that for this site, OROs would wait until the 2-5 mile area evacuation is well under way and evacuation of the 5-10 mile area begins. Shadow evacuation begins. Special facilities were notified via telephone, have mobilized resources and evacuate at this time.
4:15	Special Needs notified via telephone, have mobilized resources and evacuate at this time. Transients, who also are informed via route alerting, evacuate at this time.

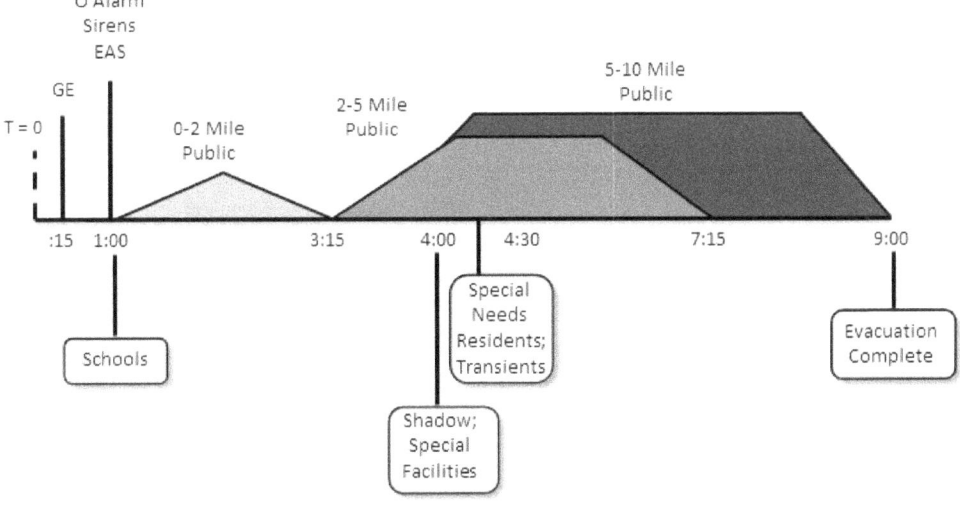

Figure 6-3. Site 2 No Siren Scenario Response Timeline

6.2 Notification Delay Scenario

In this scenario, a delay of one hour was implemented to simulate a delay in notification. The reason for the delay is not specified. The Supplement 3 response is modeled, and it is assumed that emergency response decisions and response of the public are the same as the baseline analysis, but begin one hour later. This was accomplished by setting the WinMACCS O-Alarm value at one hour. This is the only parameter that required change for this analysis. All other EP elements are unchanged from the Supplement 3 response model. The response timelines for sites 1 and 2 are described in Table 6-4 and Table 6-5 respectively and are illustrated in Figure 6-5 and Figure 6-6 respectively.

Table 6-4. Reference Site 1 STSBO Notification Delay Scenario

Time	Activity
0:00	Initiating Event
0:15	Plant declares SAE and notifies OROs. OROs initiate offsite notifications to support agencies, special facilities.
2:00	Plant declares GE and notifies OROs.
3:45 O-Alarm	OROs sound sirens and broadcast EAS message. Initial PAR is evacuation of 2 miles and SIP 2-5 miles downwind. Monitor and prepare all other areas of the EPZ. Transient 1 and Transient 2 evacuate immediately. These facilities would have been notified directly by OROs after SAE was declared. The 0-2 mile area general public begin to evacuate.
4:30	ETE90 for the 0-2 mile area is 45 minutes for this site, at which time the 2-5 mile downwind general public is instructed to evacuate. SIP is instructed for the 5-10 mile area downwind. Transient 3 evacuates
6:00	Shadow evacuation begins. By this time residents in the shadow area have observed large numbers of EPZ residents evacuating and have followed media reports covering the emergency. An actual shadow evacuation would be spread over a period of time.
6:30	After about 2 hours, begin evacuating the 5-10 mile area. At this time, evacuation of the 2-5 mile area is well underway.
8:00	Special facilities evacuate. Special facilities are modeled as evacuating at a specific time. An actual evacuation of special facilities would occur over a period of time based on mobilization needs and availability of transportation resources.
10:00	Special needs residents evacuate. Special needs residents are modeled as evacuating at a specific time. An actual evacuation of special needs residents would occur over a period of time based on mobilization needs and availability of transportation resources.

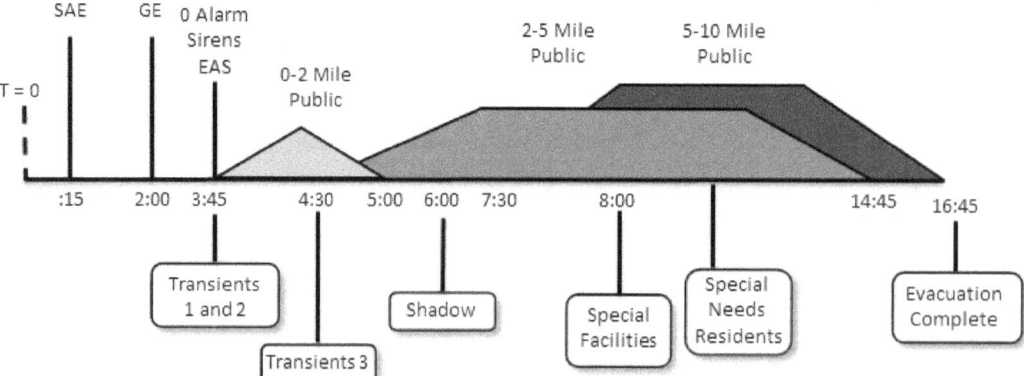

Figure 6-4. Site 1 STSBO Notification Delay Scenario Response Timeline

Table 6-5. Reference Site 2 STSBO Notification Delay Scenario

Time	Activity
0:00	Initiating Event
0:15	Plant declares immediate GE and notifies OROs.
2:00 O-Alarm	OROs sound sirens and broadcast EAS message. Protective action decision to evacuate 0-2 mile area and SIP 2-5 downwind. Evacuation of general public begins. Monitor and prepare all other areas. Schools evacuate.
3:00	2-5 mile downwind general public starts to evacuate at the ETE90 for the 0-2 mile area which is 1 hour after EAS message for this site. SIP is instructed for the 5 to 10 mile area downwind. Transients evacuate.
4:30	Shadow evacuation begins.
5:00	Assumed that after about 2 hours, begin evacuating the 5-10 mile area. At this time, evacuation of the 2-5 mile area is well underway. Special facilities evacuate.
5:30	Special needs residents begin to evacuate.

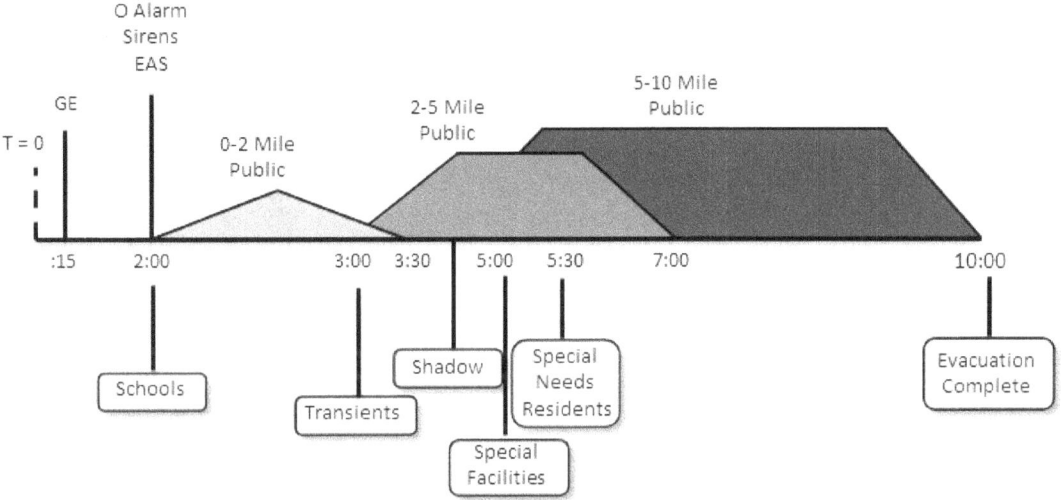

Figure 6-5. Site 2 STSBO Notification Delay Response Timeline

6.3   Analysis of Results

The results for Site 1 show that a one hour notification delay increases the dose by about 20 percent. The delay in response due to no sirens in the 2-5 mile area also shows a small increase in dose. The Site 1 results are presented in Figure 6-7.

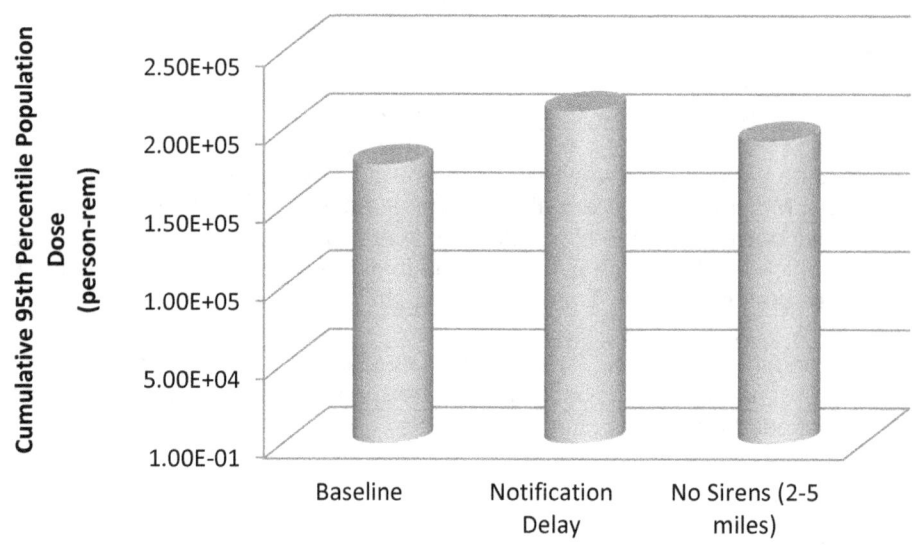

Figure 6-6. Site 1 Comparison of Emergency Planning Elements for STSBO

The results for Site 2 show that a one-hour notification delay increases the dose by more than a factor of 2. The delay in response due to no sirens in the 2-5 mile area also shows a small increase in dose. The Site 2 results are presented in Figure 6-8.

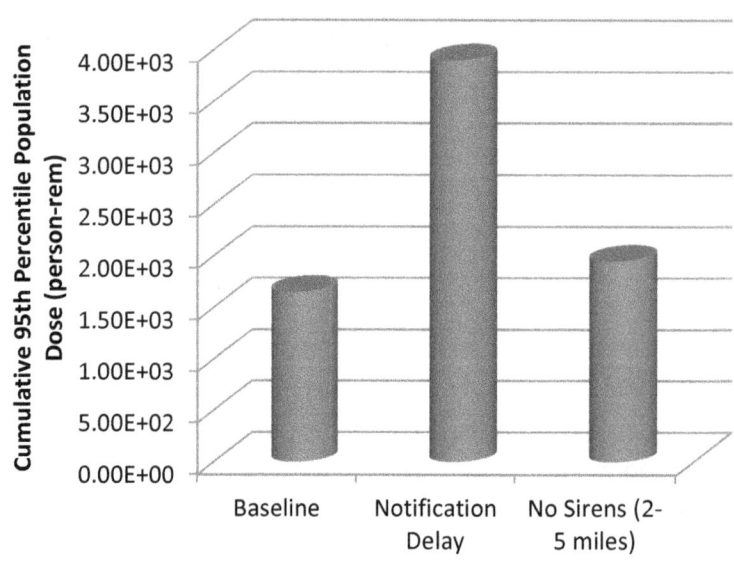

Figure 6-7. Site 2 Comparison of Emergency Planning Elements for STSBO

This analysis quantifies the importance of the time required to notify the OROs for both sites.

6.4   Uncertainty

Uncertainty exists throughout any complex analysis, and for this project, uncertainty would be found in each stage, including accident sequence selection, accident modeling using MELCOR, and consequence modeling using WinMACCS. Quantifying the value of EP using the DUQI method warrants a discussion of uncertainty. This is particularly important when the change in dose is low for a given scenario. This project used reference sites with some site specific data and some generic or default data. Because this was a demonstration effort, there was no attempt to quantify uncertainty within the analysis. The project used the latest versions of the available models and applied default or standard parameters when development of new parameters was not necessary. If this program moves forward with an actual site analysis, parameter uncertainty, model uncertainty, and completeness uncertainty should be addressed.

Each of the models used in this project incorporate many parameters. Typically, models are assumed to be appropriate. However, the parameter values for these models are often not known perfectly. Parameter uncertainties include those associated with the values of the fundamental parameters of the PRA model, such as equipment failure rates, initiating event frequencies, and human error probabilities that are used in the quantification of the accident sequence frequencies. Typically they were initially characterized by establishing probability distributions on the parameter values.

The primary models used in the analyses were MELCOR and WinMACCS. These models incorporate other models within their structure. Parameters such as speed are developed from the output of yet additional models used to analyze the evacuation. In many cases, the industry state of knowledge is incomplete, and there may be different opinions regarding how the models should be formulated. Examples include approaches to modeling human performance, common cause failures, and reactor coolant pump seal behavior upon loss of seal cooling, all of which can contribute to model uncertainty. A common approach applied to projects which are structured to provide analysis of a specific problem is to use the latest version of the models; this approach was applied to this project. The MELCOR model used the latest plant design structure for the reference plants and the latest information regarding the scenarios selected for analysis. The WinMACCS model in this project included the latest advancements related to modeling of cohorts simulating a time distribution departure of unique segments of the population and applied a 64 sector grid.

Completeness is not itself an uncertainty, but a reflection of the unanalyzed contribution. The result is, however, an uncertainty about where the true risk lies. The magnitude of completeness uncertainty is difficult to estimate. In some cases, methods of analysis have not been developed, and they have to be accepted as potential limitations of the technology. For example, the impact on actual plant risk from unanalyzed issues such as the influences of organizational performance cannot presently be explicitly assessed. For this demonstration project, use of the most current and advanced models, implemented by technical experts at each phase of the project, was considered adequate for completeness.

7.0   MITIGATION

7.1   Regulatory Concern

Nuclear power plant EP is a defense in depth measure to address uncertainty in design, construction and operational requirements. It is recognized as a cornerstone of safety and is the last barrier for protection of public health and safety during a possible, but unlikely, severe accident.

Evacuation and SIP protective actions are generally viewed as key elements for protection of public health and safety during a severe accident and this study provides a tool to analyze the efficacy of these protective actions. The emergency response system embodied in regulation, guidance and practice is intended to create a response organization capable of implementing public protective actions as well as attempting onsite actions to mitigate any accident. Thus, the licensee capability to mitigate a severe accident through maintaining core, reactor vessel and/or containment integrity should also be viewed as a key EP element.

Regulatory oversight should be designed to ensure licensees have a high level of capability to mitigate severe accidents.

7.2   Regulatory Oversight of Mitigation Capability

This study provides a tool to analyze individual EP elements associated with public protective actions, such as classification, notification of OROs, alert and notification of the public, and evacuation planning. It is possible to adapt this tool to also assess the affect of mitigation upon accident sequence. The adapted tool could be used in a performance based regulatory system to ensure that licensees develop and maintain key skills in mitigation.

The historical oversight regimen developed in 1980 includes elements of mitigation. The Technical Support Center (TSC) and Operational Support Center (OSC) are intended to analyze the accident, identify mitigative actions and implement those actions. However, the drill and exercise programs and the regulatory oversight process do not emphasize these elements. The major emphasis is placed on the "risk significant planning standards," typically communicated as "classification, notification, radiological assessment, and protective action recommendation (10 CFR 50.47(b)(4), (5), (9) and (10), respectively. Although, the oversight regimen does allow inspectors to prioritize inspection activities to include other aspects of response. Mitigation efforts are not often directly observed during drill and exercise inspections.

7.3   Background

Nuclear power plant design includes layers of procedures and installed mitigative systems to prevent core damage in the event of an off normal condition. Abnormal operating procedures, emergency operations procedures, severe accident management guides (SAMGs) and extreme damage mitigation guides (EDMGs) provide operators and the emergency response organization with direction and strategies to prevent off normal conditions from degrading and should that not be successful, to mitigate the extent of accidents. Accident mitigation is a critical component of emergency response and an effective regulatory oversight system should address the mitigative capability of a licensee.

In examining emergency response guidance beyond the nuclear industry it is interesting to note that most response guidance does not address a mitigative capability. Response phases are expressed as, "crisis" and then "response". Most industrial accidents, explosions, malicious acts and natural phenomenon are not amenable to mitigation, although response to fires is a normal mitigative action. In contrast, nuclear power plant design provides the possibility of mitigation through the use of installed equipment, containment, staged equipment and ad hoc efforts, all of which are regularly practiced during the drill and exercise program.

The overarching goal of EP is to ensure the protection of public health and safety in the case of a severe radiological accident. This is accomplished through two facets of EP: 1) implementation of protective actions such as evacuation and SIP, and 2) accident mitigation. Implementation of the DUQI method for this project illustrates the value of a formal EP program, in this case the Supplement 3 response, through comparison to a response where formal radiological emergency response planning is not established. The DUQI method could be further advanced to demonstrate the value of mitigative response. Public dose is the parameter used to measure the effectiveness of EP. However, other metrics, such as land contamination or economic cost, could be used to measure the success of mitigative efforts after evacuation is complete. The metrics may potentially be reduced if the release was contained, minimized or delayed through post core damage mitigative efforts to protect containment or delay its failure.

7.4   Current Regulatory Structure

This section discusses existing regulations and programs that are intended to enhance licensee mitigative capability. Areas where mitigation oversight might be expanded and the methods that could be employed are described below.

1. NRC requirements are assessed when the staff becomes aware of a threat not previously recognized. If it is determined that the effort will reduce risk to the public then expansion of requirements is warranted. For example, 10 CFR 50.54(hh) was added to address the risk of aircraft threat. 10 CFR 50.54(hh)(1) requires each licensee develop, implement and maintain procedures that address the following areas if the licensee is notified of a potential aircraft threat:

    - Verification of the authenticity of threat notifications;
    - Maintenance of continuous communication with threat notification sources;
    - Contacting all onsite personnel and applicable offsite response organizations;
    - Onsite actions necessary to enhance the capability of the facility to mitigate the consequences of an aircraft impact;
    - Measures to reduce visual discrimination of the site relative to its surroundings or individual buildings within the protected area;
    - Dispersal of equipment and personnel, as well as rapid entry into site protected areas for essential onsite personnel and offsite responders who are necessary to mitigate the event; and,
    - Recall of site personnel.

Inspection is planned in this area and EP related guidance recommends that this area be included at least once in drills during the exercise planning cycle. It is expected that the capability would be drilled more than once.

Oversight of this capability could be enhanced by a "mitigative response" performance indicator under the EP Cornerstone. Such an indicator would encourage licensees to conduct and critique relevant drills and provide a general assessment while minimizing direct inspection burden. However, some drills would be inspected, and the indicator itself includes burden.

Although the DUQI method was not applied in this study for mitigative response, it appears that it would be capable of determining the regulatory significance of mitigative elements. The overlapping capability of various mitigative strategies complicates the assessment, but perhaps the process should recognize redundant capability.

2. 10 CFR 50.54(hh)(2) requires each licensee develop and implement guidance and strategies intended to maintain or restore core cooling, containment, and spent fuel pool cooling capabilities under the circumstances associated with loss of large areas of the plant due to explosions or fire, to include strategies in fire fighting, operations to mitigate fuel damage, and actions to minimize radiological release.

These capabilities, called EDMGs, are inspected during a triennial fire protection inspection. One strategy is required to be demonstrated in a biennial evaluated EP exercise each planning cycle and the EP guidance recommends that all strategies (but not all variations) be drilled during an exercise planning cycle.

Oversight of this capability could be enhanced by a "mitigative response" performance indicator under the EP Cornerstone. Such an indicator would encourage licensees to conduct and critique relevant drills and provide a general assessment while minimizing direct inspection burden. However, some drills would be inspected and the indicator itself includes burden.

The regulatory significance of mitigative elements could be determined using an adaptation of the DUQI method.

3. Emergency operating procedures (EOPs) are required by plant technical specifications. These procedures provide instructions for maintaining adequate core cooling. Operators are regularly trained on EOPs and they are included in operator requalification exams. Routine simulator training on EOPs is provided to licensed operators.

NRC oversight of this capability is adequate. Operator training inspectors review this program and oversee certification of licensed operators. Demonstration of EOP implementation typically occurs during drills and exercises. The current EP inspection program does not address operator competence in EOP implementation in deference to existing regulatory oversight. However, if oversight of mitigation is to improve, inspection of EOP implementation during exercises and some drills should be achieved by including appropriate NRC expertise on exercise inspections. Additionally, oversight of this capability could be enhanced by a "mitigative response" performance indicator under the EP Cornerstone. Such an indicator would encourage licensees to conduct and critique relevant drills and provide a general assessment while minimizing direct inspection burden.

4. SAMGs are used to diagnose and mitigate a severe accident. These are operating guidelines, rather than procedures, that include steps for addressing challenges to containment integrity and reactor coolant loss beyond design basis. SAMGs are developed to enhance the capabilities of the plant emergency response team for accident sequences that progress to the point where formalized guidance may not be fully applicable (e.g., beyond the scope of emergency operating procedures). The focus is on existing plant capabilities. The primary user of the SAMGs is the TSC staff although a subset of SAMGs can be performed from the control room. SAMGs were developed with consideration of the plant specific Individual Plant Examination (IPE) evaluations.

Severe Accident Management (NRC, 1989) includes actions taken during the course of an accident to:

1. Prevent the accident from progressing to the core;
2. Terminate core damage once it begins;
3. Maintain capability of containment as long as possible; and,
4. Minimize onsite and offsite releases and their effects.

The last three are severe accident management activities after core damage has begun. Severe Accident Management Guidance (SAMG) is entered when core damage has begun.

Licensees have stated that SAMGs were verified, validated and that personnel have been trained, but maintenance of this program is not inspected. SAMGs are generally not included in exercises or drills and there is no oversight of such implementation if it does occur. Emergency plans generally do not identify positions responsible for SAMG implementation and the key skills necessary for implementation are not ensured. Use of SAMGs in drills and exercises would help to develop and maintain relevant key skills. For example, a drill might amplify the need for the emergency response organization to secure offsite support and equipment needed for containment flooding while other responders work on less drastic response options. NRC headquarters and region incident response staff are similarly not well practiced in SAMG implementation nor with supporting/understanding licensee SAMG related actions and support needs.

Oversight of this capability could be enhanced by a "mitigative response" performance indicator under the EP Cornerstone. Such an indicator would encourage licensees to conduct and critique relevant drills and provide a general assessment while minimizing direct inspection burden. However, some drills would be inspected and the indicator itself includes burden.

The regulatory significance of mitigative elements could be determined using an adaptation of the DUQI method.

7.5   Mitigation Summary

NRC oversight of mitigative response can be improved through the use of performance based and risk informed processes.

Mitigation during a severe accident requires coordination between the TSC, the Control Room and the OSC for assessing the accident, planning actions and physically carrying out those actions. The Emergency Operations Facility and the NRC Headquarters Operations Center would be involved with communication of planned actions and can assist in obtaining offsite support, if requested. The best way to develop and maintain key skills in these integrated activities is through a robust drill and exercise program with regulatory oversight. The techniques proposed above establish regulatory oversight that would enhance the protection of public health and safety.

8.0   SUMMARY AND CONCLUSIONS

The reduction in dose through the implementation of a radiological EP program provides the value of EP. In this proof of concept application, the DUQI method has shown the value of EP can be quantified. The difference in cumulative dose to the public provided the value of EP. After the value of EP was established, analyses were completed to determine whether the DUQI method was amenable to application for inidividual EP program elements. Using the STSBO accident sequences, a response was modeled considering that the EPZ siren system is not operable in the 2-5 mile area around the plant, and a response was modeled with a delay of 1 hour in the implementation of protective actions. The result presented in Table 8-1 show the cumulative population dose is reduced when implementation of a formal EP program is in place.

Table 8-1. Cumulative Population Dose for Supplement 3 and Ad Hoc Response

Sequence	Supplement 3	Ad Hoc
Site 1 STSBO	1.78×10^5	3.67×10^5
Site 1 LBLOCA	3.37×10^6	3.62×10^6
Site 2 STSBO	1.65×10^3	1.97×10^5
Site 2 ISLOCA	2.64×10^6	3.20×10^6

These results provide a metric representing the value of EP in terms of dose avoided by the public through implementation of an EP program and show that EP is amenable to being risk-informed. The DUQI method was then applied to determine risk significance of specific EP elements. Analyses were completed for evaluation of a response where sirens are assumed not operable in the 2-5 mile area around the NPP and for a delay in notification to offsite response organizations. Data for specific sites was used in selected areas to increase the validity of results, but results are not directly applicable to any specific site. The large number of cohorts and the approach to modeling for this project represents the highest fidelity use of the MACCS2 modeling code ever attempted.

The results for Sites 1 and 2 are presented in Table 8-2 and show that for Site 1, a one hour notification delay increases the dose by about 20 percent. The delay in response due to no sirens in the 2-5 mile area also shows an increase in dose, but this is not as great as the notification delay. The results for Site 2 show that a one-hour notification delay increases the dose by more than a factor of 2. The delay in response due to no sirens in the 2-5 mile area also shows an increase in dose. These results quantify the importance of the time to notify OROs.

Table 8-2. Site 1 and 2 Comparison of EP Elements to Baseline Results

Scenario	Cumulative Population Dose Site 1 (rem)	Cumulative Population Dose Site 2 (rem)
Baseline	1.78×10^5	1.65×10^3
Notification Delay	2.12×10^5	3.90×10^3
No Sirens (2-5 miles)	1.93×10^5	1.95×10^3

It is interesting to note that a delay in notification of the EPZ public could be due to untimely classification, notification, protective action recommendation development, protective action decision making or failure of equipment. This delay is more significant than a localized failure of sirens due to the effectiveness of backup notification measures, societal notification and low population density in the cases analyzed.

The objective of this study has been achieved, demonstrating through a proof of concept, that an analytical technique can be developed to risk inform EP oversight. The results of this project will allow the staff to determine whether or not it is appropriate to propose policy changes for emergency planning basis, regulations and/or guidance.

Having only studied two EP elements, no sirens and notification delay, it is important to note that quantification metric for EP elements may differ. The DUQI method provides flexibility for analysis of any EP element. The 95th percentile cumulative population dose results were used to support the study conclusions. This metric was selected based on ICRP Publication 103, (ICRP, 2007) which explains that collective dose may be used for optimization purposes for a specific range in time and space. In this analysis, only the EPZ and a seven-day emergency phase period are considered. Other metrics could be used with the DUQI method, such as early fatilities, dose thresholds, land contamination, time to release, etc., to assess value. Other criteria might also be applied, such as the number of public exposures greater than 50, 25, 5 or 1 rem.

The use of risk information can help prioritize resources while enhancing overall safety, increasing public confidence, and reducing unnecessary regulatory burden. This project has shown that EP program elements can be evaluated to determine risk significance. However, this study should be considered a proof of concept as additional cases would have to be tested and other metrics examined for usefulness before DUQI could be considered for use as a regulatory tool.

NRC oversight of mitigative response may also be improved through the use of performance based and risk informed processes. The DUQI method could be adapted for use in determining the risk significance of mitigative actions.

While the DUQI method could potentially contribute to a risk informed and performance based EP regulatory regimen, it would not be sufficient in itself. Some elements of EP programs may not be amenable to evaluation by the DUQI method. However, a performance based regulatory regimen based upon performance standards for response may be possible and the DUQI method would support such a regimen.

9.0   REFERENCES

American Society of Mechanical Engineers/American Nuclear Society (ASME/ANS). ASME/ANS RA-Sa-2009, "Addenda to ASME/ANS RA-S-2008 – Standard for Level 1/Large early Release Frequency Probabilistic Risk Assessment for Nuclear Power Plant Applications," New York, NY: ASME, 2009. (ASME/ANS, 2009).

Environmental Protection Agency (EPA). EPA-400-R-92-001, "Manual of Protective Action Guides and Protective Actions for Nuclear Incidents." Washington D.C. EPA. May 1992. (EPA, 1992).

Electric Power Research Institute (EPRI). "Risk-Informed Evaluation of Protective Action Strategies for Nuclear Plant Off-Site emergency Planning." Final Report (1015105) September, 2007. (EPRI, 2007).

ICRP Publication 103, "The 2007 Recommendations of the International Commission on Radiological Protection." (ICRP, 2007).

Nuclear Regulatory Commission (U.S.) (NRC). NUREG 0396/EPA 520/1-78-016, "Planning Basis for the Development of State and Local Government Radiological Emergency Response Plans in Support of Light Water Nuclear Power Plants." Washington D.C.: NRC. 1978. (NRC, 1978).

Nuclear Regulatory Commission (U.S.) (NRC). NUREG-0654/FEMA-REP-1, Rev. 1, "Criteria for Preparation and Evaluation of Radiological Emergency Response Plans and Preparedness in Support of Nuclear Power Plants." Washington D.C. NRC. 1980. (NRC, 1980).

Nuclear Regulatory Commission (U.S.) (NRC). Stello, Victor, Jr., 1989, "Staff Plans for Accident Management Regulatory and Research Programs", U.S. Nuclear Regulatory Commission Report SECY-89-012, Washington, D.C. (NRC, 1989).

Nuclear Regulatory Commission. NUREG 1150. "Severe Accident Risks: An Assessment for Five U.S. Nuclear Power Plants." U.S. Nuclear Regulatory Agency: Washington, DC. 1990. (NRC, 1990).

Nuclear Regulatory Commission (U.S.) (NRC). Regulatory Guide 1.174. "An Approach for Using Probabilistic Risk Assessment In Risk-Informed Decisions On Plant-Specific Changes to the Licensing Basis. Washington D.C. NRC. 1998. (NRC, 1998a).

Nuclear Regulatory Commission (U.S.) (NRC). NUREG/CR-6613, SAND97-0594. "Code Manual for MACCS2." Volumes 1 and 2. Washington D.C. NRC. 1998. (NRC, 1998b).

Nuclear Regulatory Commission (U.S.) (NRC). NUREG/CR-6864, SAND2004-5901. "Identification and Analysis of Factors Affecting Emergency Evacuations." Washington D.C. NRC. January 2005. (NRC, 2005a).

Nuclear Regulatory Commission (U.S.) (NRC). NUREG/CR-6863, SAND2004-5900. "Development of Evacuation Time Estimate Studies for Nuclear Power Plants." Washington D.C. NRC. January 2005. (NRC, 2005b).

Nuclear Regulatory Commission (U.S.) (NRC). SECY-06-0200, "Results of the Review of Emergency Preparedness Regulations and Guidance." Washington D.C. NRC. September 20, 2006. (NRC, 2006).

Nuclear Regulatory Commission (U.S.) (NRC). NUREG/CR-6953, Vol. I. SAND2007-5448P. "Review of NUREG-0654, Supplement 3, "Criteria for Protective Action Recommendations for Severe Accidents." Washington D.C. NRC. December 2007. (NRC, 2007).

Nuclear Regulatory Commission (U.S.) (NRC). NUREG/CR-6953, Vol. 2. SAND2007-5448P. "Review of NUREG-0654, Supplement 3, "Criteria for Protective Action Recommendations for Severe Accidents, Focus Groups and Telephone Survey." Washington D.C. NRC. October 2008. (NRC, 2008).

Nuclear Regulatory Commission (U.S.) (NRC). Update to Supplement 3 to NUREG-0654/FEMA-REP-1, Rev. 1, "Criteria for Protective Action Recommendations for Severe Accidents." Washington D.C. NRC. (NRC, 2011a).

Nuclear Regulatory Commission (U.S.) (NRC). NUREG/CR-7002, "Criteria for Development of Evacuation Time Estimate Studies." Washington D.C. NRC. (NRC, 2011b).

Nuclear Regulatory Commission (U.S.) (NRC). "Recommendations for Enhancing Reactor Safety in the 21st Century." Near-Term Task Force Review of Insights from the Fukushima Dai-Ichi Accident. Miller, C. L., et al. July 12, 2011. (NRC, 2011c).

Nuclear Regulatory Commission (U.S.) (NRC). NUREG/CR-7110, Volume 1. " State of the Art Reactor Consequence Analyses (SOARCA) Project: Peach Bottom Integrated Analysis." Washington D.C. NRC. January 2012. (NRC, 2012a).

Nuclear Regulatory Commission (U.S.) (NRC). NUREG/CR-7110, Volume 2. " State of the Art Reactor Consequence Analyses (SOARCA) Project: Surry Integrated Analysis." Washington D.C. NRC. January 2012. (NRC, 2012b).

RBR Consultant "Enhanced Emergency Planning. (ML092030125). December, 2007. (RBR, 2007).

Wheeler, T., Wyss, G., and Harper, F. SAND2000-2719/1. "Cassini Spacecraft Uncertainty Analysis Data and Methodology Review and Update Volume 1: Updated Parameter Uncertainty Models for the Consequence Analysis." Albuquerque, NM. November, 2000.

Wolshon, Brian, J. Jones, and F. Walton. "The Evacuation Tail and Its Effect on Evacuation Decision Making." Journal of Emergency Management. January/February 2010, Volume 8, Number 1. 201

10.0 REFERENCE REPORT

Task 3.1 Draft Letter Report: Accident Sequence Selection

Risk-Informing Regulatory Oversight of Emergency Preparedness, Identification of Representative Accident Scenarios

Task 3.1 Draft Letter Report – Rev. 2

July 27, 2011

Prepared by:

J. LaChance and J. Jones

Sandia National Laboratories
Albuquerque, NM 87185

Prepared for:

Division of Preparedness and Response
Office of Nuclear Security and Incident Response
U.S. Nuclear Regulatory Commission
Washington, DC 20555

TABLE OF CONTENTS

Page

NOMENCLATURE

ATWS	Anticipated Transient without Scram
AFW	Auxiliary Feed Water
APB	Accident Progression Bin
BWR	Boiling Water Reactor
CCW	Component Cooling Water
CDF	Core Damage Frequency
CET	Containment Event Tree
CS	Containment Spray
CSRS	Containment Spray Recirculation System
DCH	Direct Containment Heating
ECCS	Emergency Core Cooling System
EP	Emergency Preparedness
EPRI	Electric Power Research Institute
EPZ	Emergency Planning Zone
ESF	Engineered Safety Feature
FEMA	Federal Emergency Management Agency
HPI	High Pressure Injection
IPE	Individual Plant Examination
LERF	Large Early Release Frequency
LLNL	Lawrence Livermore National Laboratory
LOCA	Loss-of-Coolant Accident
LOSP	Loss of Offsite Power
LWR	Light Water Reactor
MAAP	Modular Accident Analysis Program
NEI	Nuclear Energy Institute
NPP	Nuclear Power Plant
NRC	US Nuclear Regulatory Commission
NSIR	Nuclear Security and Incident Response
ORO	Off-site Response Organization
PAI	Protective Action Instruction
PAS	Protective Action Strategy
PCS	Power Conversion System
PDS	Plant Damage State
PRA	Probabilistic Risk Assessment
PWR	Pressurized Water Reactor
RCP	Reactor Coolant Pump
RID	Representative Individual
RPV	Reactor Pressure Vessel
SAPHIRE	System Analysis Programs for Hands-on Integrated Reliability Evaluations
SBO	Station Blackout
SGTR	Steam Generator Tube Rupture
SIP	Shelter in Place
SNL	Sandia National Laboratories
SORV	Stuck-Open Relief Valve
SPAR	Standardized Plant Analysis Risk
TGE	Time of Declaration of General Emergency

1.0 INTRODUCTION

1.1 Purpose

The purpose of this project is to identify and determine whether a credible spectrum of accident scenarios can be identified for risk informing emergency preparedness (EP) requirements for existing light water reactors (LWRs). Ideally, risk-informing of EP would be accomplished by performing a full Level 3 probabilistic risk assessment (PRA) for specific plants. However, this is not currently feasible since state-of-the art Level 3 PRAs do not currently exist. Thus, an alternative approach is envisioned where knowledge from past Level 3 PRAs is combined with more current accident frequency and consequence analysis in order to identify a spectrum of severe accidents that allows risk-informed evaluation of the emergency response actions needed to protect the public. The selected accidents should include important risk contributors with credible frequencies (i.e., above a designated frequency threshold). Included are severe accidents initiated by random failures in the plant, and external hazards such as earthquakes. In addition, hostile actions against the plant are also considered even though the risk from such accidents cannot currently be evaluated. Once selected, the accident scenarios can be modeled using best estimate approaches to identify the risk-reduction potential of possible emergency response measures such as sheltering in place, staged evacuation and other measures representative of nuclear power plant (NPP) emergency response. Important uncertainties in the accident scenarios and the corresponding emergency response will be identified.

1.2 Background

Emergency preparedness is considered to be the last line of defense in the defense-in-depth philosophy. Its requirements have been established in consideration of the potential for accidents that could lead to severe core damage and the subsequent release of large amounts of radioactive material. For LWRs this release could occur in a matter of hours after the initiating event and a 10-mile plume exposure pathway Emergency Planning Zone (EPZ) has been chosen to envelope the distance beyond which it is very unlikely doses large enough to cause early fatalities would occur.

In July, 2004 Sandia National Laboratories (Sandia), working with the Nuclear Regulatory Commission's (NRC) Emergency Preparedness Directorate, began a project entitled, "Review of NUREG-0654, Supplement 3, Criteria for Protective Action Recommendations for Severe Accidents," NUREG/CR-6953 [1]. The objective of this project was to review the effectiveness of the current NRC Protective Action Recommendation (PAR) guidance contained in Supplement 3 to NUREG-0654/FEMA-REP-1 [2]. This assessment focused on whether the implementation of alternative protective actions could reduce potential health effects (i.e., early fatalities and latent cancer fatalities) in the event of an accident at an NPP. As a result of this review, the NRC staff is revising Supplement 3 to incorporate many of the recommendations of the study including staged evacuation and broader use of shelter-in-place (SIP).

Evaluation of the consequences of NPP accidents is an important aspect of risk-informing and enhancing EP. The NRC continues to examine the likely outcomes of severe reactor accidents through ongoing activities. Current activities are underway to analyze likely outcomes and provide a best estimate of the risk to the public if a severe accident were initiated at a nuclear plant. Analyses typically include scenarios with a core damage frequency (CDF) greater than 10^{-6} per reactor-year and containment bypass or early failure sequences with a CDF of greater than 10^{-7} per reactor-year. The use of a core damage frequency truncation values limits the analysis to credible yet low-frequency accident scenarios thus avoiding quantification of many scenarios that are extremely low in probability or pose only residual risk. Results of current activities cannot be used to support this project until they are published, but techniques used in these current activities have informed this project.

Industry has also performed several studies related to enhanced EP requirements. In July, 2009 the NRC staff received a technical analysis that was presented as a technical basis for enhancing EP by quantification of consequences resultant from various response actions during severe accidents. The analysis (RBR Consultants "Enhanced Emergency Planning" [3]) used certain hostile action scenarios as bounding cases for emergency response. These scenarios involved rapid releases that would be large early releases (LER) for high population sites. Using these scenarios, RBR developed a tool that can measure the impact of changes to response actions in terms of offsite hypothetical health consequences. The report goes further to suggest that protective actions could be modified to focus near the plant and rely solely on sheltering in place further away.

Industry has also performed a study related to risk-informing EP. In 2007, the Electric Power Research Institute (EPRI) published a report [4] on a risk-informed methodology for quantifying the relative effectiveness of various off-site protective action strategies (PASs). A major objective was to provide an updated technical basis for EP, including consideration of a risk-informed approach and quantification of the margin in the 10-mile EPZ. The evaluation used both the frequency and consequences of a selected set of accidents that represent a range of plant types and events, and radionuclide release timing and magnitude. The report concluded that a risk-informed approach for evaluating PASs and assessing the EP technical basis is feasible.

1.3   Objectives

The objective of this project is to identify and determine whether a credible spectrum of accident scenarios can be used to risk inform EP oversight. In addition, credible hostile actions are also to be identified for consideration when enhancing EP requirements. The NRC staff will evaluate the results of this project and propose appropriate policy changes, if any, to the EP planning basis.

This report addresses the first step in the evaluation of risk-informed emergency response measures – the selection of a set of credible accident scenarios for use in evaluating the potential emergency response for two reference plants. It provides criteria for selecting the accident scenarios and applies those criteria to available risk assessment information in order to identify a broad set of accident scenarios for use in risk-informing emergency preparedness requirements. It also includes estimation of the accident sequence frequencies which is necessary for risk-informing EP requirements. Characterization of possible accident scenarios initiated by hostile actions was also

2

performed to determine if additional accident scenarios are required for establishing EP requirements (the frequency of the hostile actions cannot be determined at this time).

2.0   DESCRIPTION OF APPROACH

A systematic method was applied to identify a spectrum of accidents for use in risk-informing EP requirements. The approach involves a review of available information in order to identify credible accidents that require emergency response actions for a reference boiling water reactor (BWR) and pressurized water reactor (PWR). The reference plants that were chosen are a BWR 4 with a Mark I containment, and a three-loop Westinghouse PWR with a sub-atmospheric containment.

A wide range of documents were reviewed in order to identify the needed spectrum of accidents. The documentation reviewed included existing Probabilistic Risk Assessment (PRA) documents [5], the Individual Plant Examinations, and the Standardized Plant Accident Response (SPAR) models for the two reference plants chosen for this study. In addition, broader perspectives for BWR and PWRs were obtained from review of both NRC and industry studies that generated a recommended list of accidents for use in establishing emergency response. These studies include NUREG/CR-6953 [1], EPRI – 1015105 [4] and other documents that specifically address the consequences from severe accidents in existing LWRs. These studies are discussed in Section 2.

The reviewed studies include a range of results that provide perspectives on important accident sequences at nuclear power plants. The results range from estimates of the core damage frequency (CDF) and timing for different accident sequences, the binning of these sequences into plant damage states (PDSs), the potential for containment failure (timing and failure modes) expressed as accident progression bins (APBs), and the resulting source term release categories (timing and magnitude). Source term release categories provide useful information on the spectrum of radiological release magnitudes and timing that one could expect from nuclear power plant accidents. However, the binning of multiple accident scenarios during the PRA process makes it difficult to identify what are the significant accident scenarios to potential offsite consequences. Thus, to be useful for this effort, the accident sequences that contribute to the source term categories must be identified. The review of the CDF, PDS, and APBs provides this information.

2.1   Accident Sequence Selection Criteria

The selection of a spectrum of accidents from the reviewed information requires establishment of a set of criteria. There are several possible criteria for selecting a set of credible accident scenarios for use in risk informing EP requirements. These can include probabilistic as well as deterministic components. Deterministic criteria include the timing and magnitude of potential radionuclide releases and the impact of the accident on evacuation (e.g., the effect of an earthquake on evacuation). For example, only accidents that result in early releases could be the focus of the assessment. This section presents a survey of possible criteria for use in this study and the selected criterion. Use of the selected criterion is discussed further in Section 4.

Probabilistic criteria can be used to eliminate scenarios that are not credible (i.e., that do not have a credible frequency of occurrence) even though they may result in significant releases. Frequency criteria can be established to address the frequency of accidents initiating events, the frequency of accident sequences resulting in core damage, the frequency of PDSs resulting in similar severe accident behavior, and the frequency of

4

radioactive release. Random, internal initiating events which are very low in frequency can be eliminated from consideration. Similarly, extremely unlikely external hazards can also be eliminated (e.g., aircraft crashes at most sites and meteorite strikes). Typically, an initiating event frequency truncation value of 1E-7/yr has been used in PRAs to eliminate initiating events from consideration. The ASME/ANS PRA Standard [6] indicates that this is an acceptable screening value as long as the event does not include an interfacing system LOCA (ISLOCA), containment bypass, or reactor vessel rupture. A low value such as this ensures that accidents and hazards that may have little or no accident mitigation potential (e.g., a vessel rupture or inadvertent airplane crash) are considered.

There are many safety systems in NPPs designed to mitigate accident scenarios. In addition, some non-safety systems are also available for accident mitigation. Although an accident initiator or hazard may have a relatively high frequency of occurrence, the availability of these mitigating systems can reduce the potential for core damage and radioactive release. PRAs are used to evaluate the potential for failure of mitigating systems following accident initiating events that result in an undesired accident end state such as core damage or radioactive material release. Level 1 PRAs evaluate the potential for core damage and Level 2 PRAs extend the analysis to the evaluation of radioactive release. Most existing PRAs are Level 1 PRAs and thus only evaluate CDF and large early release frequency (LERF) because these are two metrics used in current risk-informed regulatory applications. A CDF value of 1E-6/yr and a LERF value of 1E-7/yr are used by the NRC in regulatory guidance such as Regulatory Guide 1.174 [7] as a threshold for non-significant changes with respect to CDF and LERF, respectively.

An ongoing NRC study chose to use a CDF value of 1E-6/yr as a screening value for selecting accident sequence groups (groups of accident sequences having similar severe accident progression characteristics and timing) for inclusion in that study. In addition, a lower screening criterion of 1E-7/yr was selected for containment bypass scenarios that may have the potential for higher consequences. The 1E-7/yr screening value was used in the EPRI-1015105 EP study [4] to select the accident sequences for risk-informing EP requirements.

Other countries also utilize both CDF and a radioactive release criterion [8] for NPPs. A review of these criteria was performed to inform the selection of the criterion used in this study. With regard to CDF, most countries including the U.S utilize an upper bound CDF criterion of 1E-5/yr or 1E-4/yr for existing NPPs (some countries utilize 1E-4/yr for existing NPPs and 1E-5/yr for new NPPs). The radioactive release criteria utilized in other countries have a larger variation in the parameters used to measure a release and the associated frequency limits. Both large releases and large early releases are utilized as well as conditional containment failure probabilities are utilized. The frequency range for releases is broader than for CDF ranging from 1E-7/yr for two countries to 1E-5/yr for the majority of the countries (a 1E-6/yr criterion for new NPPs is utilized in many countries including the U.S. In general, all countries aim at using a full scope (i.e., internal and external events, at-power and shutdown modes) PRA to assess the CDF and release frequency. It is not currently known whether lower CDF and release criteria are utilized in these countries to define a threshold for non-significant contributors but it is anticipated values of at least 1% of the above criteria are reasonable threshold values for defining significant accident sequences.

The U.S. does not widely utilize risk criteria in other industries. However, some countries utilize risk criteria for use in other, non-nuclear applications. Risk guidelines are specified with regard to individuals or the society at large. Individual risk reflects the frequency that an average person located at a certain location is harmed. Generally, *individual risk* is evaluated for the most exposed individual who can be a person at an actual location or a person assumed to be constantly at the facility boundary. Characterization of the population surrounding a facility is thus not required to evaluate individual risk. *Societal risk* reflects the relationship between the frequency (F) and the number (N) of people harmed and is usually expressed in the form of an FN curve. The slope of the FN curve is defined by a *risk aversion factor* that is designed to reflect the society's aversion to single accidents with multiple fatalities as opposed to several accidents with few fatalities. Evaluation of societal risk requires determination of the population surrounding a facility. FN curves used in some European countries are shown in Figure 2-1.

The principle of <u>A</u>s <u>L</u>ow <u>A</u>s <u>R</u>easonably <u>P</u>racticable (ALARP) is utilized in many countries as an approach for achieving risk acceptance criteria in society. It is based on the following assumptions: a) There are no zero risk situations; b) Managing risk to a reasonable level is achievable; c) Acceptable risk represents the level below which an investment should be made to further reduce risk via cost-benefit analysis; d) Acceptable risk represents the *minimum risk* level that must be *obtained, regardless of cost*; e). The ALARP principle is that the *residual* risk should be <u>A</u>s <u>L</u>ow <u>A</u>s <u>R</u>easonably <u>P</u>racticable – risk reducing measures are feasible and their costs are not larger than the benefits. The principle of ALARP as applied for evaluating individual risk is illustrated in Figure 2-2. The ALARP principle as applied to societal risk is shown in Figure 2-3.

Risk acceptance criteria for individual and societal risk, though de facto exist everywhere, are not always obvious. In some Western European countries they are incorporated into law. Table 2-1 shows individual risk criteria in terms of early fatalities being used in several European countries. As indicated in the Table, a lower risk criteria ranging from 1E-8/yr to 1E-6/yr is being utilized. With regard to societal risk (see Figure 2-1), risk aversion factors of either 1 or 2 are typically utilized but with different pivot points. The acceptable criteria for a large number of early fatalities (i.e., >100) is typically less than 1E-6/yr (the exception being the United Kingdom).

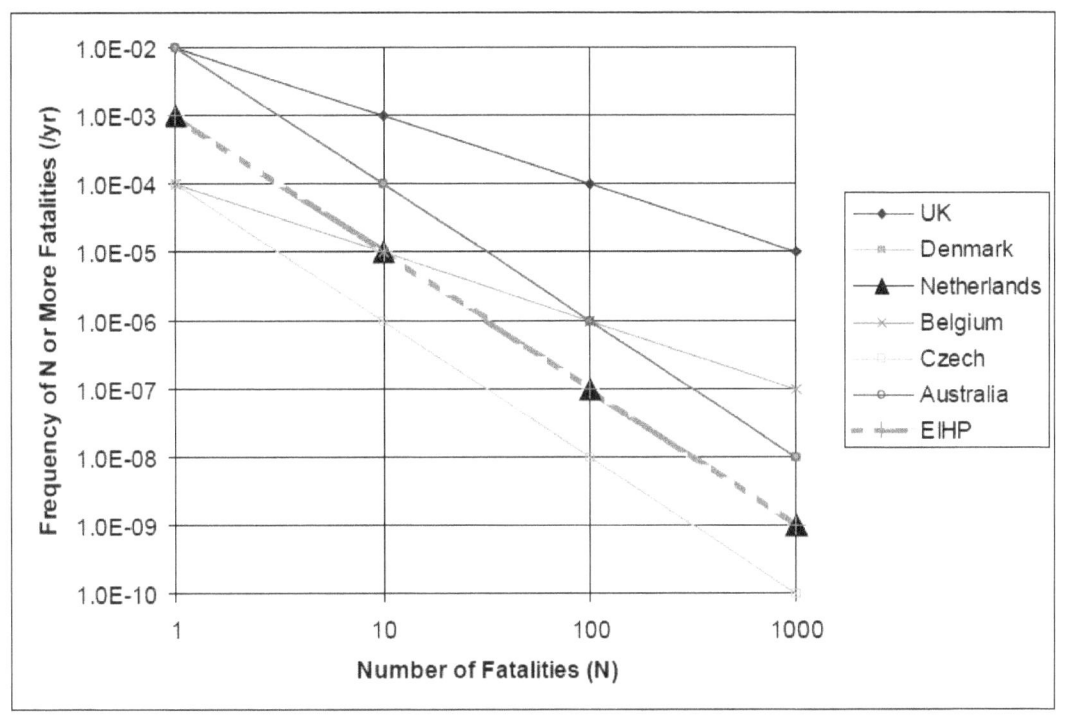

Figure 2-1. FN curves used in some European countries

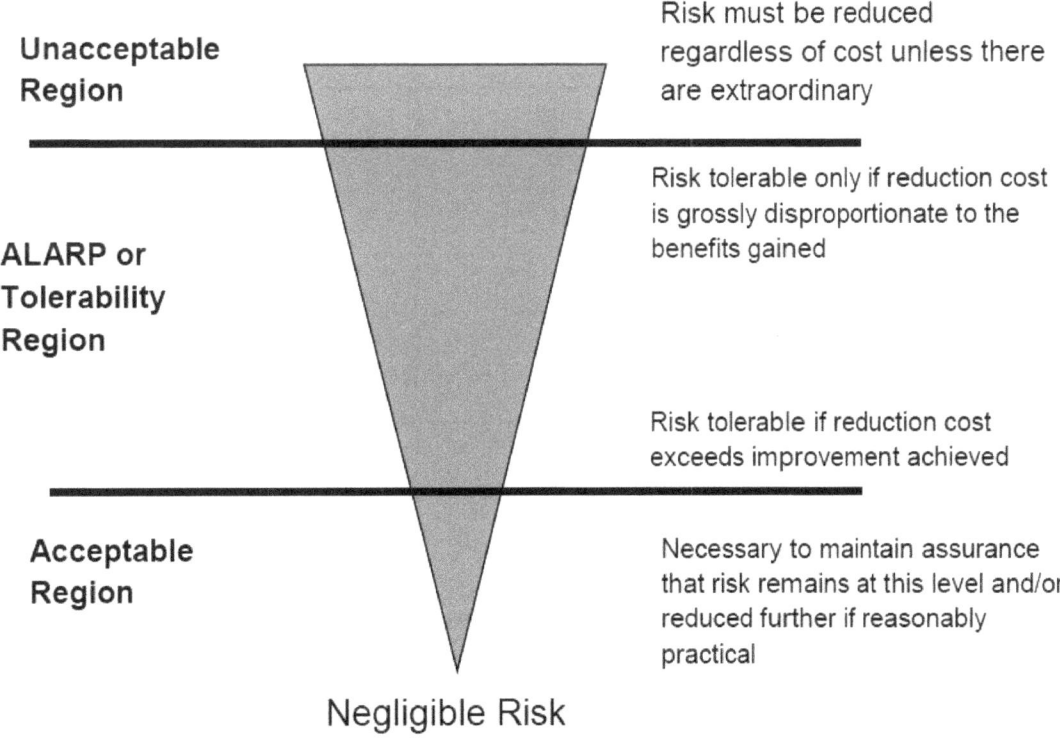

Figure 2-2. ALARP principle for individual risk

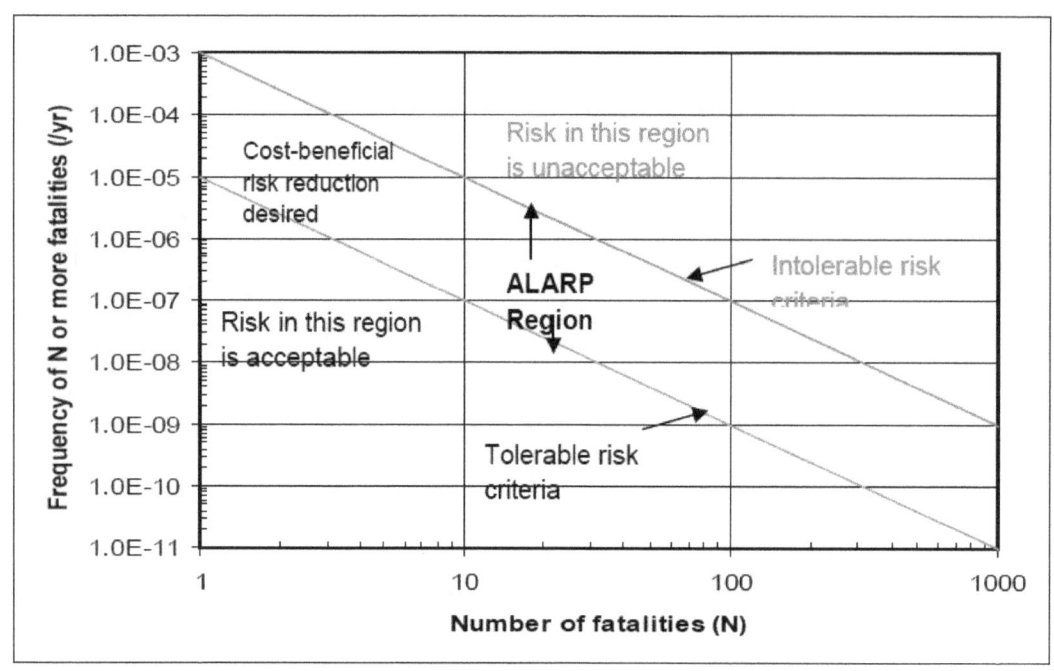

Figure 2-3. ALARP Principle illustrated for societal risk

Table 2-1. Survey of individual risk criteria for members of the public

Individual Risk Criteria	United Kingdom	The Netherlands	Hungary	Czech Republic	Australia
10^{-4}	Intolerable limit for members of the public				
10^{-5}	Risk has to be lowered to ALARP	Limit for existing nstallations, ALARA principal applies	Upper imit	Limit for existing installations, risk reduction applied.	Limit for new installations
10^{-6}	Broadly acceptable risk level	Limit for new nstallations and general limit after 2010, ALARA principal applies	Lower imit	Limit for new installations	
10^{-7}	Negligible level of risk				Negligible level of risk
10^{-8}		Negligible level of risk			

Based on the review provided above, a 1E-7/yr criterion is recommended for all levels of accident delineation (core damage scenarios to accident progression bin frequencies). This relatively low criterion is equal to or below most criteria currently in use in the U.S and abroad and is recommended for use in eliminating accident scenario types from consideration in evaluating EP requirements. This criterion has been utilized in the screening process of accident scenarios (i.e., groups of similar accident sequences) that is documented in Section 3 of this report.

3.0 Accident Scenario Review

Accident scenarios can be initiated by random failures in the plant, external hazards such as earthquakes, and by hostile actions (e.g., internal sabotage). These events can occur while the plant is at-power or when the plant is shutdown and being refueled. The magnitude of the radioactive release, the timing of the release, and the potential for affecting emergency evacuation can be different for these different scenarios. Thus, in order to provide a credible spectrum of accident scenarios for use in emergency planning, accident scenarios from different plant operating states and hazards are identified. Since the purpose of this effort is to demonstrate the feasibility of risk-informing EP requirements, the focus is on selecting possible accidents for the two selected reference plants that are being used in the feasibility study.

3.1 Probabilistic Risk Assessment Insights

Internal hazards include random failures in the plant that through some mechanism automatically trips the NPP or requires a manual shutdown. Typically, internal hazards include common transients such as turbine trips, internal fire and flood events, and loss-of offsite power (LOSP) events. External hazards include hazards that are originate offsite but impact the plant (e.g., a nearby chemical facility or inadvertent airplane crash) and natural phenomena such as earthquakes, external floods, tornados, and hurricanes. The term "at-power" refers to the normal condition when the plant is generating power. The normal at-power state is when the plant is generating 100% of the rated power. However, conditions where the plant is at low power and connected to the power grid are also possible but generally only for relatively short periods of time.
This section identifies accident scenarios from both internal and external hazards for two reference sites during both at-power and low power operation. Information was extracted from PRAs for the plants, the SPAR models for the plants, the current NRC assessments, and the plant IPEs. Some of this information is dated and does not reflect significant improvements in PRA technology. Thus, the quality of the PRA information must be considered when selecting a spectrum of accidents for use in this study. This was accomplished by weighting more current studies that have utilized more recent methodologies higher than older studies.

3.1.1 At-Power Insights - NUREG-1150

The most comprehensive at-power PRA performed on the reference sites of interest was documented in NUREG-1150 [5]. These PRAs were full Level 3 PRAs and thus evaluated the core damage frequency, containment failure probabilities and the risk to the public from internal initiators, internal fires, and some external hazards (seismic events). Although the NUREG-1150 risk assessments are outdated and do not represent the current state-of-the art in PRA, a review of the results provides a broad perspective on the type of accidents that can occur at these plants. Thus identification of the significant accident sequences from these studies provides useful insights when coupled with more up-to-date, but limited, assessments.

Table 3-1 presents a summary of the core damage frequency for different accident sequence types for both reference sites from the NUREG-1150 study. As indicated, none of the mean frequencies for the accident sequence types are less than 1E-7/yr and thus cannot be eliminated from further consideration. The dominant internal event contributors to the CDF for Site 1 are short and long-term SBO sequences involving either immediate loss of AFW (core damage occurs within 1 hour) or battery depletion (core damage occurs at approximately 7.5 hours). The dominant internal fire scenarios for Site 1 involve a loss of high pressure injection (HPI) and component cooling water (CCW) resulting in a reactor coolant pump (RCP) seal LOCA. The dominant seismic-induced scenarios involve a loss-of-offsite power (LOSP) in conjunction with either a loss of auxiliary feed water (AFW) and feed and bleed or failure of HPI and CCW leading to an RCP seal LOCA.

For Site 2, the dominant internal event contributors to CDF are SBO and ATWS scenarios. The SBO sequences are either short-term scenarios involving DC bus failures (core damage occurs within 1 hour) or long-term scenarios, with and without stuck-open relief valves (SORVs), involving battery depletion (core damage occurs between 10 to 13 hours). The ATWS scenarios include some short-term sequences (core damage occurs in approximately 15 to 20 minutes). The dominant internal fire sequences involve either a SBO or complete loss of coolant injection. The main seismic-induced contributors to core damage are SBO and a large recirculation line LOCA with an SBO.

The core damage sequences in NUREG-1150 were combined into plant damage states (PDSs) for evaluation of accident progression and containment response. Each PDS is intended to represent a unique set of circumstances with regard to the timing and conditions when core damage occurs. PDS were further grouped into coarser sets called PDS groups for propagation through the accident progression event tree. Table 3-2 presents the PDS groups from NUREG-1150 for both reference plants. Only a few PDS groups have frequencies less than 1E-7/yr and can be eliminated from further consideration based on frequency. Table 3-2 indicates that for Site 1, both short- and long-term SBO, and bypass scenarios (interfacing LOCA and SGTR) are important contributors to early containment failure (i.e., the product of the frequency of core damage and early containment failure probability exceed 1E-7/yr). For Site 2, SBO, transients with SORVs, and some ATWS sequences result in core damage and early containment failure with frequencies greater than 1E-7/yr. Fires and seismic events in both plants are important contributors to core damage and early containment failure.

The accident progression bins (APBs) from the NUREG-1150 studies are provided in Table 3-3. Only those APBs resulting in containment failure with a frequency greater than 1E-7/yr have been included. The important APBs for Site 1 include SBOs (initiated by random failures or earthquakes), bypass sequences (ISLOCAs and SGTR), and some LOCA, ATWS, and LPSD sequences. The time of release ranges from 1 to 36 hours. The number of APBs with frequencies greater than 1E-7/yr for Site 2 is substantially less and includes SBO and fire scenarios.

11

Table 3-1. Summary of core damage frequencies for Reference Sites from NUREG-1150

Accident Sequence Type	5%	Median	Mean	95%
Reference Site 1				
Internal Events	6.8E-6	2.3E-5	4E-5	1.3E-4
Station Blackout (SBO)				
Short Term	1.1E-7	1.7E-6	5.4E-6	2.3E-5
Long Term	6.1E-7	8.2E-6	2.2E-5	9.5E-5
Anticipated Transient without SCRAM (ATWS)	3.2E-8	4.2E-7	1.6E-6	5.9E-6
Transient	7.2E-8	6.9E-7	2.0E-6	6.0E-6
Loss-of Coolant Accident (LOCA)	1.2E-6	3.8E-6	6.0E-6	1.6E-5
Interfacing LOCA	3.8E-11	4.9E-8	1.6E-6	5.3E-6
Steam Generator Tube Rupture (SGTR)	1.2E-7	7.4E-7	1.8E-6	6.0E-6
Internal Fire	5.4E-7	8.3E-6	1.1E-5	3.8E-5
External Hazards				
Seismic (LLNL Hazard curves)	3.9E-7	1.5E-5	1.2E-4	4.4E-4
Seismic (EPRI hazard curves)	3.0E-7	6.1E-6	2.5E-5	1.0E-4
Reference Site 2				
Internal Events	3.5E-7	1.9E-6	4.5E-6	1.3E-5
Station Blackout	8.3E-8	6.2E-7	2.2E-6	6.0E-6
ATWS	3.1E-8	4.4E-7	1.9E-6	6.6E-6
Transient	6.1E-10	1.9E-8	1.4E-7	4.7E-7
LOCA	2.5E-9	4.4E-8	2.6E-7	7.8E-7
Internal Fire	1.1E-6	1.2E-5	2.0E-5	6.4E-5
External Hazards				
Seismic (LLNL Hazard curves)	5.3E-8	4.4E-6	7.7E-5	2.7E-4
Seismic (EPRI hazard curves)	2.3E-8	7.1E-7	3.1E-6	1.3E-5

Table 3-3 provides timing information important to EP including the assumed warning time, time at which the radioactive release to the environment begins, the assumed time evacuation begins, and the release duration. The warning time represents the time at which a general site emergency is declared and emergency actions including evacuation are initiated. The warning times for Site 2 used in NUREG-1150 corresponds to the time at which the coolant level falls below two feet above the bottom of the active fuel. For Site 1, the warning time definition is more variable but generally reflects the time at which the operators have a clear indication that a core melt is imminent or in progress. The warning times range from 0 to 6 hours for Site 1 and 1 to 8 hours for Site 2. The majority of the Site 1 scenarios assumed warning times of 6 hours but a few LPSD scenarios utilized 0 warning times and ISLOCAs assumed warning times of 22 minutes.

The time of release ranges from 1 to 36 hours with most releases occurring between 8 and 13 hours (the 1 hour releases occur for the LPSD and ISLOCA sequences). The release time for Site 2 sequences range from 3.6 to 11.4 hours. The time evacuation begins was generally shortly after the warning time and before the beginning of the release. Significant delays in evacuation were assumed for seismic sequences. The duration of the releases range from 3 to 24 hours.

Table 3-2. Plant Damage State Group Frequencies from NUREG/CR-4550

PDS Group	Description	Mean Core Damage Frequency (/yr)	Mean Early Containment Failure Probability
Site 1			
Internal Initiators			
PDS-1	Slow (Long-Term) SBO	2.2E-5	8.0E-3
PDS-2	LOCA	6.1E-6	6.0E-3
PDS-3	Fast (Short-Term) SBO	5.4E-6	7.0E-3
PDS-4	Interfacing LOCA	1.6E-6	NA
PDS-5	Transient	1.8E-6	2.0E-3
PDS-6	ATWS	1.4E-6	3.0E-3
PDS-7	SGTR	1.8E-6	NA
Fire	Fire results in RCP seal LOCA with no ECCS	1.1E-5	1.8E-2
Seismic			
EQ-1 (LLNL)	LOSP (no SBO)	9.1E-5	0.1
EQ-2 (LLNL)	SBO	7.9E-5	
EQ-3 (LLNL)	LOCAs	2.3E-5	
EQ-1 (EPRI)	LOSP (no SBO)	1.5E-5	Not Calculated
EQ-2 (EPRI)	SBO	9.4E-6	
EQ-3 (EPRI)	LOCAs	3.0E-6	
Site 2			
Internal Initiators			
PDS-1	LOCA – ECCS Injection Failure	2.6E-7	0.38
PDS-2	Transient with two SORVs – ECCS Failure	2.2E-7	0.51
PDS-3	Transient with two SORVs – ECCS Failure	6.1E-9	
PDS-4	Fast (short-term) SBO – No DC (HPI fails, ADS fails)	2.1E-7	0.60
PDS-5	Slow (long-term) SBO – Battery	1.9E-6	

Table 3-2. Plant Damage State Group Frequencies from NUREG/CR-4550

PDS Group	Description	Mean Core Damage Frequency (/yr)	Mean Early Containment Failure Probability
	Depletion		
PDS-6	Fast ATWS – HPI fails, LPI available	3.0E-7	
PDS-7	ATWS – IORV, SLC fails	1.1E-7	
PDS-8	ATWS - SLC fails	1.5E-6	0.51
PDS-9	ATWS – LOSP, LPI available	4.4E-8	
Fire			
PDS-1	Fast Transient	6.8E-6	0.3
PDS-2	Slow SBO	5.9E-6	0.9
PDS-3	Slow SBO	5.7E-6	0.9
PDS-4	Long Transient	1.1E-6	0.8
Seismic (LLNL)			
PDS-1	LOSP with RPV failure	8.9E-6	1.0
PDS-2	Fast SBO, Large LOCA	1.7E-5	1.0
PDS-3	Fast SBO, Large LOCA	3.0E-6	1.0
PDS-4	Slow SBO	3.7E-5	0.8
PDS-5	Fast SBO	3.2E-6	0.7
PDS-6	Fast SBO, ISLOCA	4.7E-6	0.9
PDS-7	Fast SBO	1.6E-6	0.6
Seismic (EPRI)			
PDS-1	LOSP with RPV failure	3.3E-7	1.0
PDS-2	Fast SBO, Large LOCA	6.3E-7	1.0
PDS-3	Fast SBO, Large LOCA	1.4E-7	1.0
PDS-4	Slow SBO	1.6E-6	0.8
PDS-5	Fast SBO	1.9E-7	0.7
PDS-6	Fast SBO, ISLOCA	1.9E-7	0.9
PDS-7	Fast SBO	7.2E-8	0.6

Table 3-3   Accident Progression Bins from NUREG-1150 studies

Accident Progression Bin	Plant Damage State Group	Containment Failure	Core Damage Frequency (1/RY)	Conditional Containment Failure Probability	Containment Failure Frequency (1/RY)	Warning Time (s)	Time of Beginning of Release (s)	Evacuation Time (hr)	Release Duration (hr)
					Site 1				
GDCDFCDBDDB	Slow (long-term) SBO	None	2.20E-05	9.09E-01	2.00E-05	2.20E+04	4.70E+04	6.94E+00	2.39E+01
GHCDFCDBDDB	Low G Earthquake - LOSP	None	1.40E-05	7.54E-01	1.06E-05	2.20E+04	4.70E+04	6.94E+00	2.39E+01
GHADBCAADDB	Fire	None	1.10E-05	6.90E-01	7.59E-06	2.20E+04	4.70E+04	6.94E+00	2.39E+01
GDCDFCDADDB	LOCA	None	6.10E-06	9.38E-01	5.72E-06	2.20E+04	4.70E+04	6.94E+00	2.39E+01
GDCDFCDBDDB	Fast (short-term) SBO	None	5.40E-06	9.09E-01	4.91E-06	2.20E+04	4.70E+04	6.94E+00	2.39E+01
GHADBCABDDB	Low G Low G Earthquake - SBO	None	8.40E-06	5.46E-01	4.59E-06	2.20E+04	4.70E+04	6.94E+00	2.39E+01
FHADBCABDCB	Low G Earthquake - LOSP	Late	1.40E-05	2.32E-01	3.25E-06	2.20E+04	1.30E+05	3.00E+01	3.06E+00
FHADBCAADCA	Fire	Late	1.10E-05	2.92E-01	3.21E-06	2.20E+04	1.30E+05	3.00E+01	3.06E+00
FHADBCABDCB	Low G Low G Earthquake - SBO	Late	8.40E-06	3.82E-01	3.21E-06	2.20E+04	1.30E+05	3.00E+01	3.06E+00
CHADBCABDBAB	LP&SD	Early	4.22E-06	6.46E-01	2.73E-06	0.00E+00	3.60E+03	1.00E+00	6.00E+00
GHADBBAADDA	SGTR	None	1.80E-06	1.00E+00	1.80E-06	3.60E+04	5.10E+04	4.17E+00	6.39E+00
GDCDFCDBDDB	Transient	None	1.80E-06	9.79E-01	1.76E-06	2.20E+04	4.70E+04	6.94E+00	2.39E+01
BHADBCAADDA	ISLOCA	Wet	1.60E-06	1.00E+00	1.60E-06	1.30E+03	3.70E+03	6.67E-01	6.61E+00
AHADBCAADEA	ISLOCA	Dry	1.60E-06	1.00E+00	1.60E-06	1.30E+03	3.70E+03	6.67E-01	6.61E+00
GHEDBCABDDAA	LP&SD	None	4.22E-06	3.27E-01	1.38E-06	0.00E+00	3.60E+03	1.00E+00	6.00E+00
GHCDFCDADFB	Low G Earthquake - LOCA	None	2.50E-06	5.34E-01	1.34E-06	2.20E+04	4.70E+04	6.94E+00	2.39E+01
GDCDFCDBDDB	ATWS	None	1.40E-06	8.73E-01	1.22E-06	2.20E+04	4.70E+04	6.94E+00	2.39E+01
GHCDFCDBDFB	High G Earthquake -LOSP	None	1.10E-06	7.54E-01	8.29E-07	2.20E+04	4.70E+04	6.94E+00	2.39E+01
CHADBCAADCB	Low G Earthquake - LOCA	Early	2.50E-06	3.29E-01	8.23E-07	2.20E+04	2.80E+04	1.67E+00	6.39E+00
GHADBCABDFB	High G High G Earthquake - SBO	None	1.10E-06	5.46E-01	6.01E-07	2.20E+04	4.70E+04	6.94E+00	2.39E+01
CHADBCAADCB	Low G Low G Earthquake - SBO	Early	8.40E-06	7.00E-02	5.88E-07	2.20E+04	2.80E+04	1.67E+00	6.39E+00

Table 3-3   Accident Progression Bins from NUREG-1150 studies (continued)

Accident Progression Bin	Plant Damage State Group	Containment Failure	Core Damage Frequency (1/RY)	Conditional Containment Failure Probability	Containment Failure Frequency (1/RY)	Warning Time (s)	Time of Beginning of Release (s)	Evacuation Time (hr)	Release Duration (hr)
GHCDFCDADDB	High G Earthquake - LOCA	None	9.80E-07	5.34E-01	5.23E-07	2.20E+04	4.70E+04	6.94E+00	2.39E+01
FFADBCABDCB	Fast (short-term) SBO	Late	5.40E-06	7.90E-02	4.27E-07	2.20E+04	1.30E+05	3.00E+01	3.06E+00
FHADBCABDCB	High G High G Earthquake - SBO	Late	1.10E-06	3.82E-01	4.20E-07	2.20E+04	1.30E+05	3.00E+01	3.06E+00
FHADBCAADCB	Low G Earthquake - LOCA	Late	2.50E-06	1.37E-01	3.43E-07	2.20E+04	1.30E+05	3.00E+01	3.06E+00
FDDDBCAADCB	LOCA	Late	6.10E-06	5.50E-02	3.36E-07	2.20E+04	1.30E+05	3.00E+01	3.06E+00
CHADBCAADCB	High G High G Earthquake -LOCA	Early	9.80E-07	3.29E-01	3.22E-07	2.20E+04	2.80E+04	1.67E+00	6.39E+00
FHADBCABDCB	High G Earthquake - LOSP	Late	1.10E-06	2.32E-01	2.55E-07	2.20E+04	1.30E+05	3.00E+01	3.06E+00
DHACCCAABCA	Fire	Early	1.10E-05	1.80E-02	1.98E-07	2.20E+04	2.80E+04	1.67E+00	6.39E+00
DHADDCBADBB	Slow SBO	Early	2.20E-05	8.00E-03	1.76E-07	2.20E+04	2.80E+04	1.67E+00	6.17E+00
DHACACABACB	Low G Earthquake - LOSP	Early	1.40E-05	1.20E-02	1.68E-07	2.20E+04	2.80E+04	1.67E+00	6.39E+00
FHADBCAADCB	High G Earthquake - LOCA	Late	9.80E-07	1.37E-01	1.34E-07	2.20E+04	1.30E+05	3.00E+01	3.06E+00
EFADBCABDAAA	LP&SD	Late	4.22E-06	2.60E-02	1.10E-07	0.00E+00	3.60E+03	1.00E+00	6.00E+00
					Site 2				
GAABFBAAAA	Fire - Slow (long-term) SBO	Early	6.90E-06	8.73E-01	6.02E-06	2.90E+04	4.10E+04	3.33E+00	4.14E+00
GAABFBAAAA	Fire - Slow (long-term) SBO	Early	6.02E-06	8.55E-01	5.15E-06	2.90E+04	4.10E+04	3.33E+00	4.14E+00
BBEEICDCAA	Fire - Fast Transient	None	5.94E-06	5.97E-01	3.55E-06	4.00E+03	2.20E+04	5.00E+00	6.36E+00
BBDEFBBBAA	Fire - Fast Transient	Early	5.94E-06	3.29E-01	1.95E-06	4.00E+03	1.30E+04	2.50E+00	3.94E+00
GAABFBAAAA	Slow (long-term) SBO	Early	1.90E-06	6.26E-01	1.19E-06	2.90E+04	4.10E+04	3.33E+00	4.14E+00
GAABACAAAB	Fire Slow (long-term) SBO	Late	6.02E-06	1.39E-01	8.37E-07	2.90E+04	4.70E+04	5.00E+00	8.61E+00

Accident Progression Bin	Plant Damage State Group	Containment Failure	Core Damage Frequency (1/RY)	Conditional Containment Failure Probability	Containment Failure Frequency (1/RY)	Warning Time (s)	Time of Beginning of Release (s)	Evacuation Time (hr)	Release Duration (hr)
GAABACAAAB	Fire Slow (long-term) SBO	Late	6.90E-06	1.20E-01	8.28E-07	2.90E+04	4.70E+04	5.00E+00	8.61E+00
GAABFBAAAB	Fire Transient CV	Early	9.42E-07	8.42E-01	7.93E-07	2.90E+04	4.10E+04	3.33E+00	4.14E+00
DAABFBAAAA	ATWS CV	Early	1.40E-06	5.36E-01	7.50E-07	1.70E+04	2.60E+04	2.50E+00	4.14E+00
GAABFBAAAA	High G Seismic Slow SBO	Early	6.30E-07	8.55E-01	5.39E-07	2.90E+04	4.10E+04	3.33E+00	4.14E+00
AABDFBAACA	High G Seismic FSB LLOCA	Early	5.00E-07	1.00E+00	5.00E-07	4.00E+03	1.30E+04	2.50E+00	3.94E+00
GBEEICDCAA	Slow SBO	None	1.90E-06	2.33E-01	4.43E-07	2.90E+04	4.70E+04	5.00E+00	6.36E+00
CBEEICDCAA	ATWS CV	None	1.40E-06	2.42E-01	3.39E-07	4.00E+03	2.20E+04	5.00E+00	6.36E+00
AABDFBAACA	High G Seismic FSB RPV	Early	2.50E-07	1.00E+00	2.50E-07	4.00E+03	1.30E+04	2.50E+00	3.94E+00
CBDEFBBBAA	Fast ATWS	Early	3.50E-07	5.36E-01	1.88E-07	4.00E+03	1.30E+04	2.50E+00	3.94E+00
AABDFBAACA	High G Seismic FSB ILOCA	Early	1.50E-07	9.30E-01	1.40E-07	4.00E+03	1.30E+04	2.50E+00	3.94E+00
BGEEGCDCAB	Slow SBO	Late	1.90E-06	6.80E-02	1.29E-07	2.90E+04	4.70E+04	5.00E+00	4.14E+00
EBDEFBBBAA	Fast SBO	Early	2.00E-07	6.26E-01	1.25E-07	4.00E+03	1.30E+04	2.50E+00	3.94E+00
AABDFBAACA	High G Seismic FSB LLOCA	Early	1.20E-07	1.00E+00	1.20E-07	4.00E+03	1.30E+04	2.50E+00	3.94E+00

3.1.2 SPAR Model Results

The NRC sponsors the development and maintenance of plant-specific PRAs for every commercial nuclear power plant in the U. S. These PRAs have been constructed in a relatively consistent manner under the Standardized Plant Analysis Risk (SPAR) program and currently only include internal events. In the past, these internal event SPAR models have been limited to the estimation of risk at the core damage frequency level (known as a Level-1 PRA). In recent years, the NRC has sponsored the development of few SPAR models that estimate the risk of a release of radioactive material into the environment (i.e., Level-2 PRAs). The approach taken by the NRC was to fund the expansion of some existing Level-1 SPAR models to support the development Level-2 models by including the various containment systems at NPPs that affect the response of the containment structure (and subsequent likelihood of a radioactive release), but do not significantly affect the likelihood of core damage (hence, are not included in original Level-1 SPAR models). This work was performed at the Idaho National Laboratory (INL) and included "extended" Level-1 models for both reference sites. These models have been generated taking into account features included in recent licensee PRAs for these plants (i.e., updates of the plant IPE models). As such, they better reflect the current understanding of the risk contributors for these plants than are reflected in the NUREG-1150 and IPEs assessments.

The finished integrated Level-1/Level-2 SPAR model for the reference sites consists of the following features:

- **Level-1 extended event trees:** These are the original SPAR level-1 event trees that have been modified to include the plant containment systems needed for modeling the response of the containment structure to the core damage accident sequence.

- **Plant Damage State event trees:** Instead of terminating at core damage, in the Level-2 analysis the accident sequences are extended to identify the response of the plant to the severe accident and predict the likelihood of a radioactive release to the environment. The core damage sequences are binned into plant damage states to facilitate this effort. A PDS event tree containing important accident sequence characteristics is used to accomplish this binning.

- **Containment event tree:** The containment event tree (CET) tracks the progression of the severe accident, from the onset of core damage through the challenges to the reactor pressure vessel (RPV) and the containment structure. Each PDS is propagated through the CET resulting in various containment responses to the severe accident sequences and subsequent releases of radioactive material. A separate source term category event tree is used to bin the CET sequences into source term release categories.

Table 3-4 presents the PDSs obtained from solution of the reference site Level 2 SPAR models that have CDFs greater than 1E-7/yr. Important PDSs for Site 1 include most of the types of accident scenarios (SBO, transients, ATWS, LOCAs, ISLOCAs, and SGTRs). The Site 2 PDSs are more limited and do not include SBOs or ISLOCAs. The importance of SBOs at Site 2 has decreased significantly compared to the NUREG-1150 study due to credit for tying into a downstream dam.

20

Table 3-4. Important PDSs from SPAR model evaluation

PDS	PDS Vector	PDS Description	Frequency
Site 1			
35	PDS-XNTNZLAMANZ	Transient - long term containment heat removal (CHR) failure	7.21E-07
50	PDS-XNSNZZNMAAN	Medium/small LOCA - failure of secondary heat removal (SHR) and coolant injection	6.91E-07
62	PDS-XLZNZZZHZNZ	SGTR - large early release	5.34E-07
6	PDS-XNTBENNHAAN	SBO - power recovered before RPV failure but coolant injection fails	4.48E-07
61	PDS-XIZNZZZHZNZ	ISLOCA	3.39E-07
57	PDS-XNANZZNHAAD	ATWS - failure of SHR and coolant injection	1.49E-07
14	PDS-XNTBELNMAAN	SBO - with SORV or RCP seal LOCA, failure of SHR and coolant injection	1.35E-07
54	PDS-XNANZZAHAAN	ATWS - failure of coolant injection	1.06E-07
42	PDS-XNLNZZZLAAN	Large LOCA - failure of coolant injection	1.00E-07
Site 2			
6	PDS-XNTZHZSZSSF	Transient - RPV at high pressure, no coolant injection	3.12E-07
2	PDS-XNTZLZFVSFF	Transient - RPV at low pressure, no coolant injection, containment vented	1.56E-07
23	PDS-XNAZHZZZFFF	ATWS - no coolant injection	1.36E-07
5	PDS-XNTZLZFFFFF	Transient - RPV at low pressure, no coolant injection	1.05E-07
37	PDS-XNRZLNFVSSF	SORV- RPV at low pressure, no coolant injection, containment vented	1.03E-07
14	PDS-XNLZLNFFFFF	Large LOCA - no coolant injection	8.51E-08

3.2   Current NRC Activities

The NRC has current activities underway to develop a better understanding of the realistic outcomes of severe accidents in existing LWRs. The severe accident modeling incorporates significant plant improvements not reflected in earlier assessments such as NUREG-1150. Improvements in systems, training and emergency procedures, offsite emergency response, and recent security-related enhancements have occurred that can affect the risk from severe accidents. In addition, there have been improvements in the state-of-the art in modeling severe accidents behavior and evaluation of consequences to the public.

Ideally, risk-significant sequences could be identified by reviewing the results of a full-scope Level 3 PRA. Unfortunately, there are few full-scope Level 3 PRAs and those that exist do not reflect the improvements discussed above (e.g., the NUREG-1150 studies discussed in Section 3.1). However, there are many Level 1 PRAs for internal events that can be utilized to identify dominant core damage sequences. Current NRC activities reviewed for this project utilize CDF information combined with an understanding of containment loads and failure mechanisms during severe accidents to select the accident sequences for evaluation, and thus have elected to analyze sequences with CDFs greater than 1E-6/year. For sequence groups involving containment bypass, sequence with CDFs greater than 1E-7/year were selected for analysis because of the potential for these sequences to have higher consequences and higher risk. To accomplish this, the release characteristics were grouped so that they are representative of scenarios binned into those groups and the groups are sufficiently broad to include the potentially risk-significant but lower-frequency scenarios.

Core-damage sequences from previous NRC and licensee PRAs were identified and binned into core-damage groups. A core-damage group consists of core-damage sequences that have similar characteristics with respect to severe accident progression (timing of important phenomena) and containment or engineered safety feature operability. The groups were screened according to their approximate core-damage frequencies to identify the most significant groups. Finally, the accident scenario descriptions were augmented by assessing the status of containment systems (which are not typically modeled in Level 1 PRAs).

The scenarios generated by internal events and the availability of containment systems for these scenarios were identified using the NRC's plant-specific SPAR models, licensee PRAs, and other risk information sources. The following process was used in another current NRC activity to determine the scenarios for further analyses:

1.   Candidate accident scenarios were identified in analyses using plant-specific, SPAR models (Version 3.31).

     a. Initial Screening – Screened out initiating events with low CDFs (<1E-7) and sequences with a CDF <1E-8. This step eliminated 7% of the overall CDF for Site 1 and 4% of the overall CDF for Site 2.
     b. Sequence Evaluation – Identified and evaluated the dominant cutsets for the remaining sequences. Determined system and equipment availabilities and accident sequence timing.
     c. Scenario Grouping – Grouped sequences with similar times to core damage and equipment availabilities into scenarios.

2.   Containment systems availabilities for each scenario were assessed using system dependency tables which delineate the support systems required for performance of the target front-line systems and from a review of existing SPAR model system fault trees.

3.   Core-damage sequences from the licensee PRA model were reviewed and compared with the scenarios determined by using the SPAR models. Differences were resolved during meetings with licensee staff.

4.   The screening criteria (CDF < 10^{-6} for most scenarios, and < 10^{-7} for containment bypass sequences) were applied to eliminate scenarios from further analyses.

Detailed sequence characteristics are more difficult to specify for scenarios initiated by external hazards (e.g., fire, seismic, flooding) due to the lack of external event PRA models industry-wide. The external event scenarios selected for analysis in the project are representative of those that might arise due to seismic, fire or internal flooding initiators. Although they were derived from a review of past studies such as the NUREG-1150, individual plant examination for external event (IPEEE) submittals, and other relevant generic information, they do not represent specific accident sequences from any of these prior studies.

In order to specify the scenarios for further analysis and the assessment of mitigation measures, the selected scenarios were assumed to be seismically initiated since in general, seismic-initiated scenarios are the most restrictive in terms of the ability to successfully implement onsite mitigation measures and offsite protective actions. In addition, the seismic-initiated scenarios were judged to be important contributors to the external event core damage and release frequencies.

Current NRC activities have included assessment of the scenarios presented in Table 3-5 for the reference site plants.

3.3   Industry Studies

Three industry documents related to emergency planning were reviewed as part of the effort for selecting the accident sequences for use in risk-informing EP. The first is an EPRI study under NEI co-sponsorship [4] that addresses the same subject matter - risk-informing protective action strategies for NPP off-site emergency planning. As such, a review of the accident sequences selected for that study provides input to this effort and also identifies if the spectrum of accidents in the EPRI study is sufficient for risk-informing EP. The second document is a white paper also prepared by EPRI and NEI on the evaluation of accident scenario timing for emergency planning of "fast breaking events" [9]. This white paper addresses the response to accident sequences that can result in early radiological releases. The identified scenarios in this white paper are a potential subset of accidents that would be included in the spectrum of accidents needed to risk-inform EP. The third document, produced by RBR Consultants [3] also provides a technical basis for risk informing EP by quantifying the consequences associated with various response actions during severe accidents. All three reports are summarized in the following subsections.

Table 3-5. Scenarios Selected for Consequence Analysis in Current NRC Activities

Scenario	Initiating Event	Representative CDF (PRY)	Description of Scenario
Site 1			
Long-term SBO	Seismic, fire, flooding	2E-5	Immediate loss of ac power and eventual loss of control of turbine-driven systems due to battery depletion
Short-term SBO	Seismic, fire, flood	2E-6	Immediate loss of ac power and turbine-driven systems
TISGTR	Seismic, fire, flood	4E-7	Immediate loss of ac power and turbine-driven systems, consequential tube rupture
Interfacing systems LOCA	Random failure of check valves	3E-8	Check valves in high-pressure piping fail open causing low-pressure piping outside containment to rupture, followed by operator error
Site 2			
Long-term SBO	Seismic, fire, flood	3E-6	Immediate loss of ac power and eventual loss of control of turbine-driven systems due to battery exhaustion
Short-term SBO	Seismic, fire, flood	3E-7	Immediate loss of ac power and turbine-driven systems

3.3.1 EPRI Risk-Informed Evaluation of Protective Strategies

The EPRI risk-informed EP study [4] integrates improvements in our knowledge of severe accidents with emergency planning experience to evaluate potential protective action strategies. The objectives of this project as stated in Reference 4 are:

- "To develop a risk-informed (R-I) methodology for quantifying the relative effectiveness of various off-site PASs. Depending on the effectiveness and practicality of the implementation by the off-site response organization (ORO) and the public, these strategies could then be considered for use in the emergency planning (EP) process for nuclear power plants."

- "To provide an updated technical basis for EP, including consideration of an R-I approach and quantification of the margin in the 10-mile emergency planning zone (EPZ) required in the regulations."

The EPRI study selected a generic set of severe accident sequences and associated source terms for use in the development and demonstration of their risk-informed EP process. The report indicates that the selected source terms are applicable to a broad range of accidents sequences and plant types. A sequence frequency threshold of 1E-7 per year was selected as a reasonable bound for including accident sequences resulting in significant radiological releases. However, the report states that the risk assessment was done both with and without the 1E-7/yr frequency truncation value (i.e., both for accident sequences greater than 1E-7/yr and for all sequences regardless of frequency).

The accident sequence and source term selection process invoked by EPRI followed the same general approach as is being pursued in this study –a review of existing PRA information. The EPRI review focused on PWR internal events and included some information not reviewed in this study:

- NUREG 1150 PWR results for Surry, Zion, and Sequoyah[10,11,12]
- PWR IPEs for the three NUREG 1150 plants [13,14,15,16]
- A recent industry study on the timing of severe accidents [17]
- Recent industry work on risk from induced steam generator tube ruptures (SGTRs) [18,19,20]
- An EPRI study on the consequences of bypass accidents [21]
- NUREG 0654, the NRC regulatory guidance for emergency response [2]

Based on the review of the above sources, seven core damage accident sequence types were defined for use in the EPRI study. Table 3-6 lists the accident types and the values of some key parameters including the time of declaration of a general emergency (TGE). As indicated in the table, the seven accidents represent a wide range of accident sequence frequencies, release timing, and release magnitude. Mean values from NUREG-1150 were used for most of the listed parameters since they were believed by the EPRI report authors to be the most appropriate and conservative compared to IPE results. For parameters where a range of values exist, which occurred when results from both the IPE and NUEG-1150 for a plant were utilized, central values were generally selected with greater weight given to values from more recent or detailed work. As indicated in Table 3-6, two of the seven accident sequence types evaluated in the study have frequencies less than 1E-7/yr. The risk evaluation of different PASs performed in

24

the study are based on the sequences with frequencies greater than 1E-7/yr although results for all the sequences were included as additional information.

Information on other accident sequence types was also reviewed in the EPRI study to confirm that the source terms resulting from the PWR, internal events-related information provide reasonable representations of these accident types, which include BWR internal event accidents [22,23], PWR and BWR fire-initiated and seismic accidents [10,23], and terrorist-initiated accidents [24,25]. Based on the comparison of the information in Table 3-6 with that on other accident sequence types, the EPRI report concluded that the Table 3-6 source terms are representative for BWR and PWR plants, for internal and external events, and for terrorist-initiated events.

Table 3-6. List of PWR Accident Sequences from EPRI 1015105

Accident Sequence Type	Frequency(yr^{-1})	Beginning of Release (hours after scram)	TGE (hours after scram)	Iodine Release Fraction*
1. LOCA early containment failure	5E-7	3	1	0.1
2. Fast SBO early containment failure	3E-7	4.5	1	0.15
3. Spontaneous SGTR	2E-6	16	7	0.2
4. Induced SGTR	5E-9	3	1.5	0.2
5. ISLOCA	3E-8	4	1	0.25
6. LOCA auxiliary bldg release	5E-6	3	1	0.01
7. Core damage sequence: intact containment	5E-5	5	4	1E-5

3.3.2 EPRI White Paper on "Fast-Breaking Events"

The EPRI (Polestar) white paper was generated in response to proposed criteria from the Federal Emergency Management Agency (FEMA) for evaluating the capability of OROs to respond to "fast-breaking events." This concept evolved over concerns associated with potential terrorist activity against nuclear power plants. The response to these "fast-breaking events" would require accelerated response that could essentially bypass the Emergency Action Levels (EALs) that are in place to address events at nuclear power plants. The EPRI white paper examines a range of accident sequences for various plant types that could lead to early radioactive releases in order to determine if there is a need for special consideration of "fast breaking accidents" in emergency response planning. Thus, the joint EPRI and NEI white paper is useful for this effort in that it provides useful information on types of accidents that may be challenging to emergency response due to the short time frame for potential release of radioactive material. Any such accidents are important to consider in risk-informing EP.

"Fast breaking events" are defined in the white paper primarily on the basis of the timing of resulting significant radioactive releases. Three factors were used in this classification:

1. The first notification of the OROs of a problem at the plant comes in the form of a general emergency (or is so close after a less severe notification as to be effectively the first indication of a problem).
2. Core damage and release of fission products from the fuel occurs rapidly following the initiation of the scenario (within the first hour after the initiating event).
3. Containment failure is occurring or imminent at the time of notification of the OROs (within 1 hour of the release of significant fission products from the fuel).

In addition, accident sequence frequencies were also considered in defining "fast breaking events." Based on a review of guidance and precedents for use of accident frequencies in risk-informed decisions, 1E-7 per year was selected as a reasonable bound for including accident sequences in the "fast-breaking event" classification.

The white paper dismissed accident scenarios involving boil-off of the reactor coolant as "fast breaking events" on the basis that core damage will generally occur between 2 to 3 hours after initiation of the event, vessel breach will occur later between 6 to 7 hours, and the likely containment failure mode is overpressurization which would occur after 20 hours (although it was recognized that containment failure could occur at vessel breach). These time frames were considered to be too long to be "fast breaking events."

Loss of reactor coolant inventory accidents or LOCAs involving both loss of emergency coolant during either the injection or recirculation phase were also considered in the EPRI white paper. Recirculation phase failures were dismissed as "fast breaking events" since recirculation failures generally occur hours after the initiation of the LOCA. The white paper thus only considered LOCAs with an early loss of emergency coolant injection since they tend to result in core damage quicker than loss of heat removal sequences. The scenarios reviewed are listed in Table 3-7. External events and terrorist attacks that could lead to early core uncovery were also considered. The timing of the potential releases for the scenarios in Table 3-7 was determined using the Modular Accident Analysis Program (MAAP 4.0.4).

The conclusions reached from the EPRI evaluation are that the FEMA-proposed requirements for "fast breaking events" are not necessary and actually could negatively impact public health and safety by exposing the general public to a process that does not allow the ORO sufficient time to properly consider all the factors important to the emergency response. It was concluded that external events and terrorist attacks would result in similar accidents to those given in Table 3-7 with similar timing of core damage and containment failure. In addition, it was concluded that it would be unlikely that an accident initiated by a terrorist attack would completely bypass the entire EAL system. The white paper further states that existing emergency response requirements are adequate for dealing with potential core damage accidents that could result in radioactive material release. The potential for a significant offsite release from a nuclear power plant accident would not begin for a minimum of several hours.

Table 3-7. Accident Sequences from EPRI (Polestar) White Paper

Sequence Type	Mitigation Failures	Time to Beginning of Core Damage	Time to Beginning of Significant. Offsite Release
PWR Sequences			
Small LOCA	Loss of injection (ECCS)	1.1 to 1.4 hrs	3 to >24 hours
Medium LOCA	Loss of injection (ECCS)	0.8 to 1.0 hrs	3 to >24 hours
Large LOCA	Loss of injection (ECCS)	0.2 to 0.3 hrs	~24 hrs or greater
Spontaneous SGTR	Loss of injection and isolation	>16 hrs	>16 hrs
Induced SGTR	High pressure core damage and ruptured tube	3 to 12 hrs	3 to 12 hrs
ATWS	Loss of reactivity control and secondary heat removal	See comments	See comments
ISLOCA	Loss of low pressure injection and recirculation	3 to 6 hrs	3 to 6 hrs
Inventory loss (shutdown)	Loss of injection (ECCS)	>3 hr	>3 hr
BWR Sequences			
Small LOCA	Loss of injection (ECCS)	0.6 hr	~4 hrs
Medium LOCA	Loss of injection (ECCS)	0.4 hrs	~4 hrs
Large LOCA	Loss of injection (ECCS)	0.2 hrs	~4 hrs
ATWS	Loss of reactivity control and level control	~1.5 hr	~1.5 hr
Inventory loss (shutdown)	Loss of injection (ECCS)	>3 hr	>3 hr

3.3.3 RBR Enhanced Emergency Planning Study

The purpose of the RBR report [3] was "to support efforts to enhance emergency planning and to suggest fundamental principles for a new emergency planning paradigm." The technical analysis documented in this report presents a technical basis for risk informing EP based on the quantification of consequences from bounding severe accidents in a specified PWR. Using these scenarios as input, RBR utilized a newly developed tool that can measure the impact of changes to response actions in terms of offsite hypothetical health consequences. Based on the results, the report concludes that protective actions could be modified to focus near the plant and rely solely on sheltering in place further away. Note that since this report is listed as proprietary information, detailed information on these results is not provided here.

The RBR analysis was based on two severe accident scenarios assumed to be initiated by a terrorist attack. Both scenarios assume a successful terrorist attack which results in breaching the containment within 30 minutes. The onset of core damage was assumed to occur immediately after containment failure due to terrorist destruction of engineered safety systems. The first scenario is an SBO scenario where all sources of offsite and onsite electrical power were assumed disabled and other non-electrical means to cool the core (i.e., turbine-driven auxiliary feedwater) was assumed inoperable at some point. In the second scenario, the terrorist are assumed to have severed a major primary cooling water pipe resulting in a large break LOCA and to also disable all emergency core cooling systems (ECCS) designed to respond to this type of accident. In addition, the containment spray systems were also assumed to be disabled thus eliminating an important mechanism for reducing the amount of radioactive material that might be released to the environment.

Although both of these scenarios involve early core damage and containment failure, the time of radioactive material release does not occur for several hours due to the additional time required for the core melt to penetrate the reactor vessel which supports the EPRI white paper (see Section 3.3.2) conclusion that there are no "fast breaking scenarios". This additional time is important for initiating emergency response actions. Although the containment is assumed to be open and the containment spray systems are inoperable, a large fraction of iodine and other radioactive isotopes that are important contributors to early health effects are retained within the containment. The duration of the release ranges from 13 to 14 hours for these two scenarios with the release rate decreasing rapidly. Information on the timing of these two sequences and the fraction of material released is provided in Table 3-8.

Table 3-8. Summary of release characteristics from RBR scenario evaluation

Scenario	Time of Release (hr)	Fraction of Iodine Released	Fraction of Cesium Released	Fraction of Tellurium Released
SBO	4.4	0.274	0.180	0.182
LOCA	2.0	0.111	0.101	0.121

It is noted that the 2 hour time for release for the LOCA scenario does not agree with the results of MELCOR analyses performed by SNL to determine source terms for high burnup cores (see Section 3.4).

3.4   Source Term

An important perspective to consider in selecting a set of scenarios for risk-informing EP is the different source terms that can be generated. Both the magnitude of the radioactive material release and timing is important. To provide this perspective, several references related to source terms were reviewed. The first is NUREG-1465 [26] which presents a source term that can be applied to the design of light water reactors. The developed source term is based on a range of severe accidents that have been analyzed for existing LWRs. The work in NUREG-1465 has recently been expanded by SNL to examine the source terms in LWRs that utilize high burnup cores [27,28]. The SNL studies utilized advances in the understanding of severe accident progression and fission product release and transportation to generate best estimate analyses of selected accident sequences. Of particular interest to this review is the fact that in addition to evaluating severe accidents for high burnup cores (i.e., greater than 40 GWD/MTU), the response in existing burnup cores was also evaluated. Thus, these reports provide information on a set of severe accidents that should be analyzed in addition to information on the calculated response. It is important to note that the generated source terms in these reports are "in-containment source terms" and not source terms released to the environment. This is because the focus of these efforts was to provide source terms for evaluating compliance with 10 CFR 100 [29] requirements, which is based on containment leakage and not containment failure. Additional regulatory applications of this source term include post-accident equipment qualification and post-accident control room habitability assessment.

The accidents considered in generating the NUREG-1465 source terms are provided in Table 3-9. Accidents from the reference plants as well as other LWRs were included in this assessment. The evaluation of the range of the severe accidents in Table 3-9 is based upon the work done in NUREG-1150 and involves complete core melt, failure of the reactor vessel, and core-concrete interactions. Table 3-9 also provides information on the risk significance of the selected accidents based on information from the Individual Plant Examination (IPE) insights report NUREG-1560 [30].

As with the NUREG-1465 accident sequences, the accident sequences analyzed in the SNL high burnup core source term assessments are meant to reflect a representative set of severe accident scenarios. They do not necessarily include all risk-significant accidents. The selection of accident sequences evaluated in the SNL study utilized the information provided in Table 3-9. The selected calculation matrix covered the range of accidents included in the NUREG-1465 evaluation and considers insights from NUREG-1560. Unfortunately, resource limitations did not allow for evaluation of all of these accident sequences. However, it was judged that a reasonable set of accident analyses could be performed by modeling the sequences listed for Surry, Sequoyah, Peach Bottom, and Grand Gulf. This is primarily due to similarities in accident sequences across plant types. The selected accidents are provided in Table 3-10. Additional information on the resulting source terms (timing of release and fraction of radionuclides released) is provided in Reference 27 and 28.

29

Table 3-9. Sequences used in NUREG-1465 assessment

Sequence	Description	Risk Significance (NUREG-1560)	Other Comments
Surry			
AG	LOCA (hot leg), no containment heat removal systems	Moderate	Large LOCAs currently are thought to be minimally risk significant
TMLB'	LOSP, no power conversion system (PCS)	High	
V	Interfacing system LOCA	Low	IPE identification of potential bypass path led to operator training to minimize risk
S3B	SBO with RCP seal LOCA	High	
S2D-δ	Small break LOCA, no ECCS and H_2 combustion	High	
S2D-β	Small break LOCA with 6" hole in containment	Not discussed	
Peach Bottom			
TC1	Anticipated Transient without Scram (ATWS), reactor depressurized	Low	
TC2	ATWS, reactor pressurized	Low	
TC3	ATWS, reactor pressurized, wetwell vented	Low	
TB1	SBO, battery depletion	High	
TB2	Same as TB1 except CF at VF	High	Pressure @VF or shell melt through
S2E1	2" equivalent diameter LOCA, no ECCS, no ADS (high pressure)	Low	
S2E2	Same as S2E1 except PB concrete replaced with basaltic concrete	Low	
V	Residual Heat Removal (RHR) system pipe failure outside containment	Low	
TBUX	SBO, loss of all DC power	High	
LaSalle			
TB	SBO with late containment failure		
Grand Gulf			
TC	ATWS, early containment failure fails ECCS		
TB1	SBO with battery depletion		
TB2	TB1 with H_2 burn failing containment		
TBS	SBO, no ECCS but reactor depressurized		
TBR	TBS with AC power recovered after vessel failure		
Zion			

Table 3-9. Sequences used in NUREG-1465 assessment

Sequence	Description	Risk Significance (NUREG-1560)	Other Comments
S2DCR	LOCA (2"), no ECCS and no containment spray recirculation system (CSRS)	High	
S2DCF1	RCP seal LOCA, no ECCS, containment sprays (CS), and fan coolers – H2 burn or direct containment heating (DCH) fails containment	Low	Early failure unlikely for large dry containment. Vessel pressure reduced by LOCA, prevents high pressure melt ejection at vessel failure
S2DCF2	S2DCF1 except late H_2 overpressure fails containment	High	
TMLU	Transient, no PCS, ECCS, auxiliary feedwater system – DCH fails containment	High	Current thinking is containment failure at vessel failure is less likely since primary is likely depressurized.
Oconee 3			
TMLB'	SBO, no active emergency safeguard feature (ESF) systems	High	
S1DCF	LOCA (3"), no ESF systems	Moderate	
Sequoyah			
S3HF1	RCP seal LOCA, no ECCS, no CSRS – reactor cavity flooded	High	
S3HF2	S3HF1 with hot leg-induced LOCA	Not discussed	
3HF3	S3HF1 with dry reactor cavity	Not discussed	
S3B	LOCA (0.5") with SBO	Low	
TBA	SBO induces hot leg LOCA – H_2 burn fails containment	High	
ACD	LOCA (hot leg), no ECCS, no CS	Moderate	
S3B1	SBO results in delayed RCP	High	
S3HF	RCP seal LOCA, no ECCS, no CSRS	High	
S3H	RCP seal LOCA, no ECCS recirculation	High	

Table 3-10. Sequences Analyzed in SNL High Burnup Study

Case	Description
Surry	
1A	SBO, no ECCS and AFW, RCP seal failure, late containment failure (47 hours)
1B	Small LOCA, no ECCS, AFW and CS operates, late containment failure
1C	Large LOCA, ECCS and CS injection, late containment failure (>168 hrs)
1D	SBO, no ECCS and AFW, and no RCP seal failure, late containment failure
1F	Small LOCA, no ECCS, AFW operates, late containment failure at vessel breach (21.7 hours)
Peach Bottom	

31

Table 3-10. Sequences Analyzed in SNL High Burnup Study

Case	Description
1A	Short-term SBO, SORV and no coolant injection, early containment failure (drywell liner melt-through at 9.51 hours)
1D	Short-term SBO, vessel at high pressure, no coolant injection, early containment failure (drywell head flange leaks at 10.5 hours)
1B	Short-term SBO, SORV and no coolant injection, early containment failure (drywell liner melt-through at 9.6 hours), core-concrete interaction included
1C	Short-term SBO, SORV and no coolant injection, late containment failure (drywell liner melt through at 9.5 hours)
2A	Long-term SBO (8 hrs), SORV, early containment failure (drywell liner melt-through at 24.5 hours)
2B	Long-term SBO (8 hrs), SORV, late containment failure (drywell head flange leakage at 25.3 hours)
2C	Long-term SBO (8 hrs), SORV, late containment failure (torus over pressurization at 28.9 hours)
3	Small LOCA (steam line), early containment failure (drywell head flange leakage at 8.8 hours)
4	Small LOCA (steam line), early containment failure (drywell melt through at 7 hours)
Sequoyah	
4A	RCP seal LOCA in 1 loop, no ECCS, AFW and CS available, cavity flooded, containment failure (78.5 hours)
4B	RCP seal LOCA in 1 loop, no ECCS, AFW and CS available, cavity not flooded, containment failure (90 hours)
4C	RCP seal LOCA in 1 loop, ECCS, AFW and CS available, containment failure (91.1 hours)
4D	Short-term SBO, no ECCS, steam-driven AFW available for 1 hour, containment failure (87.3 hours)
4E	Short-term SBO, no ECCS and AFW, early containment failure (6.3 hours)
4F	Large LOCA, no ECCS, late containment failure (41.8 hours)
4G	Small LOCA, no ECCS and AFW, late containment failure (62.8 hours)
Grand Gulf	
5A	Short-term SBO, SORV and no coolant injection, early containment failure (H_2 burn at vessel breach results in containment failure at 10.5 hours)
5B	Short-term SBO, no coolant injection, early containment failure (H_2 burn at vessel breach results in containment failure at 8.7 hours)
5C	Short-term SBO, SORV and no coolant injection, late containment failure (overpressure failure at 64 hours)
6A	Long-term SBO (8 hrs), SORV, early containment failure (H_2 burn at vessel breach results in containment failure at 17.7 hours)
6B	Long-term SBO (8 hrs), SORV, late containment failure (overpressure failure at 57.1 hours)
7	ATWS, coolant injection available, containment failure at 8.2 hours prior to core damage
8	Large LOCA, only RCIC available, late containment failure (overpressure failure at 36 hours)

4.0 Accident Scenario Selection

Section 3 of this report provides a broad review of the types of accident sequences that are important with respect to various risk measures including core damage, containment failure, and source terms. The majority of the information provided is related to the two plants chosen for this study on risk-informing emergency preparedness requirements. However, some of the information provided covers a broader spectrum of plants. This section provides the results of the effort to condense this information into a set of accident sequences for use in risk-informing EP requirements. Section 4.1 provides the criteria that were used in this effort and Section 4.2 applies the criteria and provides a recommended set of accident sequences.

4.1   Selection Criteria

The goal of this effort is to identify a set of credible accident scenarios that bound the potential emergency response for two reference plants. "Credible" was defined in Section 2 as any accident sequence or accident group with a CDF greater than 1E-7/yr. In addition to this criteria, the following additional criteria have been utilized for selecting representative accident scenarios:

1. Accident sequences that can be caused by random failures, external events, or terrorist acts should be selected to reduce the number of scenarios requiring detailed evaluation.
2. Similarly, accident sequences that provide similar source terms for both PWR and BWRs and for different operating ranges (i.e., at-power versus LPSD) should be considered in order to reduce the number of scenarios requiring evaluation.
3. Although the emphasis is on selected reference plants, it is desirable that the selected accident sequences reflect the important scenarios for similar plant types.
4. Although accident scenarios identified as being important in multiple studies should be considered for inclusion, the selected scenarios should reflect the most recent information possible with regard to frequency and importance to risk. More weight should be given to recent studies and resulting information in selecting the accident sequences (e.g., the SPAR internal event model results are felt to better reflect the reference plant internal event risk than was calculated in NUREG-1150 and IPE studies).
5. More emphasis should be placed on early release sequences as they provide the most challenges to emergency response actions.
6. It is desirable to include accident sequences evaluated in similar industry risk-informed EP studies in order to compare the results and insights.
7. In order to leverage recent, state-of-the art MELCOR analyses and minimize the amount of additional analysis, it is desirable to select accident sequences that have been recently analyzed in NRC projects and the SNL high burnup core source term evaluations.

4.2   Selected Accident Sequences

Table 4-1 presents a preliminary set of accident sequences recommended for inclusion in the risk-informed EP project. The accident sequences all meet the 1E-7/yr criteria. In addition, the sequences are reflected in the results of the NUREG-1150 study, IPEs, SPAR model evaluations, industry studies, NUREG-1465 source term, the SNL high burnup core evaluations, and current NRC activities. Table 4-1 identifies whether each sequence meets the criteria identified in Section 4.1.

All of the sequences selected result in early releases and thus would provide the most significant challenge to EP actions. Long-term scenarios were considered but were eliminated as sufficient time would be available to accomplish necessary EP actions. Similarly, the recommended sequences would result in substantial releases of radionuclides because of either containment failure or bypass (scenarios involving containment leakage were not considered because of the small source term). Each of the selected sequences could be caused by multiple hazards or by hostile actions. For some of the scenarios, it is recommended that the hostile actions be assumed to change the timing of some events (e.g., the hostile action can be assumed to result in early containment failure).

Accident Scenario	Accident can be caused by multiple hazards?	Accident applicable to other plants?	Selection Criteria			
			Scenario important in recent models/ studies?	Sequence reflects early release potential?	Sequence included in industry studies	Recent MELCOR analysis of sequence?
		Site 1				
Short-term SBO, immediate loss of TDAFW, consequential SGTR (	Y	Y	Y	Y	EPRI, Polestar, RBR	Current NRC activities
Large LOCA, failure of coolant injection, early containment failure	Y	Y	Y	Y	EPRI, Polestar, RBR	HBU
		Site 2				
Short-term SBO (with SORV), failure of turbine-driven systems	Y	Y	Y	Y	EPRI	Current NRC activities, HBU
ISLOCA	Y	Y	Y	Y	EPRI	

REFERENCES

1. Jones, J.A., Bixler, N., Schelling, F.J., "Review of NUREG-0654, Supplement 3, Criteria for Protective Action Recommendations for Severe Accidents," NUREG/CR-6953, Sandia National Laboratories, Albuquerque, NM, December 2007.

2. Nuclear Regulatory Commission (U.S.) (NRC). Supplement 3 to NUREG-0654/FEMA-REP-1, Rev. 1, "Criteria for Protective Action Recommendations for Severe Accidents," Washington D.C.: NRC. 1996.

3. "Enhanced Emergency Planning," RBR Consultants, Inc., 2007 [Proprietary Information].

4. "Risk-Informed Evaluation of Protective Action Strategies for Nuclear Plant Off-Site Emergency Planning," EPRI-1015105, Electric Power Research Institute, Palo Alto, CA, September 2007.

5. "Severe Accident Risks: An Assessment for Five U.S. Nuclear Power Plants, Final Summary Report," *NUREG-1150, v. 1*, U.S. Nuclear Regulatory Commission, Washington D.C., December, 1990.

6. "Standard for Level 1/Large Early Release Frequency Probabilistic Risk Assessment for Nuclear Power Plant Applications, Addenda to ASME/ANS RA-S-2008," ASME/ANS RA-Sa-2009, American Society of Mechanical Engineers, New York, NY, February 2009.

7. "An Approach for Using Probabilistic Risk Assessment in Risk-Informed Decisions on Plant-Specific Changes to the Licensing Basis," Regulatory Guide 1.174, Revision 1, U.S. Nuclear Regulatory Commission, Washington, DC, November 2002.

8. "Probabilistic Risk Criteria and Safety Goals," NEA/CSNI/R(2009)16, Nuclear Energy Agency Committee on the Safety of Nuclear Installations, December 17, 2009.

9. Leaver, D., Li, J., Mays, S., True, D, Gaertner, J., Vine, G., "Evaluation of Accident Scenario Timing for Emergency Planning of 'Fast-Breaking Events'," *White Paper,* Nuclear Energy Institute, December, 2003.

10. Breeding, R.J., Helton, J.C, Murfin, W.B., Smith, L.N., "Evaluation of Severe Accident Risks: Surry Unit 1, Main Report," NUREG/CR-4551, Vol. 3, Rev. 1, Part 1, U.S. Nuclear Regulatory Commission, Washington D.C., October, 1990.

11. C.K. Park, et al., "Evaluation of Severe Accident Risks, Zion Unit 1, Main Report," NUREG/CR-4551, Vol. 7, Rev. 1, Part 1, U. S. Nuclear Regulatory Commission, Washington D.C, March 1993.

12. J.J Gregory, et al., "Evaluation of Severe Accident Risks, Sequoyah, Unit 1, Main Report," NUREG/CR-4551, Vol. 5, Rev. 1, Part 1, U. S. Nuclear Regulatory Commission, Washington D.C, December 1990.

13. "Probabilistic Risk Assessment for the Individual Plant Examination, Surry Power Station Units 1 and 2," Virginia Power Company, Final Report, August 1991.

14. "MACCS2 Model for Surry Level 3 Application," SM-1241, Virginia Power Company, February 2000.

15. "Zion Nuclear Generating Station, Units 1 and 2, Individual Plant Examination Submittal Report," Commonwealth Edison Company, April 1992.

16. Sequoyah Individual Plant Examination, submitted to U.S. Nuclear Regulatory Commission in August 1992.

17. D. Leaver and S. Mays, et al,. 'An Evaluation of Accident Scenario Timing for Emergency Planning of Fast Breaking Events,' White paper prepared under sponsorship of EPRI for NEI, December 2003.

18. "Steam Generator Tube Integrity Risk Assessment," TR-107623, Vol. 1, Rev. 1: General Methodology, EPRI, Palo Alto, CA: 2002.

19. E. Fuller (EPRI), "Surry Early Induced SGTR Release, Vol. 1, Rev. 1: General Methodology," E-mail message to D. Leaver (Polestar), April 22, 2006.

20. E. Fuller (EPRI), Zion Early Induced SGTR Release. E-mail message to D. Leaver (Polestar), July 5, 2006.

21. "Evaluation of the Consequences of Containment Bypass Scenarios," NP-6586-L, EPRI, Palo Alto, CA: 1989.

22. Payne, A.C., Breeding, R.J., Jow, H.-N., Helton, J.C., Smith, L.N., Shiver, A.W., "Evaluation of Severe Accident Risks: Peach Bottom, Unit 2, Main Report," NUREG/CR-4551, Vol. 4, Rev. 1, Part 1, U.S. Nuclear Regulatory Commission, Washington, DC, December, 1990.

23. T.D. Brown, et al., "Evaluation of Severe Accident Risks, Grand Gulf Unit 1, Main Report," NUREG/CR-4551, Vol. 6, Rev. 1, Part 1, U. S. Nuclear Regulatory Commission, Washington D.C December 1990.

24. "Risk Characterization of the Potential Consequences of an Armed Terrorist Ground Attack on a U.S. Nuclear Power Plant." Electric Power Research Institute, Palo Alto, CA., Limited distribution; contact Nuclear Energy Institute for information.

25. "Deterring Terrorism: Aircraft Crash Impact Analyses Demonstrate Nuclear Power Plant's Structural Strength." Electric Power Research Institute, Palo Alto, CA., Limited distribution; contact Nuclear Energy Institute for information.

26. L. Soffer, et al., "Accident Source Terms for Light-Water Nuclear Power Plants,"NUREG-1465, U. S. Nuclear Regulatory Commission, Washington D.C February 1995.

27. S. G. Ashbaugh, et al. "Accident Source Terms for Pressurized Water Reactors with High-Burnup Cores Calculated Using MELCOR 1.85," SAND2008-6664, Sandia National Laboratories, Albuquerque NM, April 2010.

28. M.T. Leonard, R.O. Gauntt, and D.A. Powers, "Accident Source Terms for Boiling Water Reactors with High-Burnup Cores Calculated Using MELCOR 1.85," Draft Report, Sandia National Laboratories, Albuquerque NM, 2006.

29. "Reactor Site Criteria," Title 10, Code of Federal Regulations, Part 100, January 2002.

30. "Individual Plant Examination Program: Perspectives on Reactor Safety and Plant Performance, NUREG-1560, Volumes 1 and 2, December 1997.

Appendix A

Review of the RBR Report

REVIEW OF THE RBR REPORT

In July, 2009 NRC received a technical report entitled "Enhanced Emergency Planning," (the RBR Report) that was presented as a technical basis for risk informing EP through the quantification of consequences resultant from various response actions during severe accidents (RBR, 2007). The RBR report used selected hostile action scenarios as bounding cases for emergency response representing LERs for a high population site. Using these scenarios, RBR developed a tool that can measure the impact of changes to response actions in terms of offsite hypothetical health consequences and demonstrated a risk based approach to enhance emergency planning.

The RBR report used a loss of coolant accident and station blackout accident initiated through hostile action to provide source terms for the consequence analysis. Breach containment was assumed at 30 minutes at which time an accident sequence was initiated. A two hour delay was assumed between reactor scram and start of the public evacuation. The core release data in the report shows the first plume segment at 2.0 hours for the loss of coolant accident and at 4.4 hours for the station blackout accident. The report explains that the loss of coolant and station blackout accidents encompassed all of the source terms of the risk significant accident scenarios for the selected site. The basis for the release timing and containment breach is not described in detail within the report and thus was not reviewed.

The RBR report provides a technically advanced approach that merged traffic analysis with the MACCS2 consequence model. This advancement allows the evacuations to be modeled as waves of people leaving at different times from different initial locations and more precisely modeled the location of the public during the evacuation. Everyone in a single wave travelled along the same route and up to three speeds were used per route. People that had not yet departed were assumed to be sheltered. The shielding values used in the report were similar to the shielding values used in this study. The RBR report assumed a daytime midweek scenario using the Indian Point EPZ (RBR, 2007) as the demonstration site. The high fidelity model was applied to a distance of four miles. The report explains that because the ranges of early fatalities and early injuries fall well within four miles, it was not necessary to make a detailed tracking of the evacuation of people who start to evacuate from locations beyond four miles to determine early health effects (RBR, 2007).

A wide range of parameters were analyzed and consequences were reported in terms of early and latent fatalities. Variations in the percent of public compliance, travel speeds, timing, distance from the plant, and sheltering were analyzed. Results were typically presented for the 95th and 100th percentile using early fatalities as the metric although early injury and latent fatalities were also presented in some cases. Benefits were shown where notification and response of the public occurs one-half hour earlier than the base case and for use of inhalation protection. The report also showed that speeding up evacuation through the use of improved traffic control (e.g., contra flow / reverse laning) reduced consequences. The timing to implement contra flow was not discussed.

The RBR report showed that early fatalities were few under most conditions and these only occurred within 4 miles of the plant. The report concludes that these early fatalities could be reduced if residents beyond 4 miles were to shelter allowing those nearer the plant to evacuate more quickly. The report provided a general evaluation of emergency planning showing results for 'No Emergency Response,' 'Minimum Emergency Response,' and 'Basic Emergency Response' with each increase in the level of response reducing the number of latent fatalities.

The report provides 15 recommendations, some of which are described below:

- A goal of 100 percent public participation should be established.
- Family emergency planning should be encouraged.
- Surveys should be conducted of the public that live within 2 miles to ensure they have a means to evacuate.
- Transit dependent strategies within 2 miles of the plant should be reconsidered.
- Schools within 4 miles should work closely with response agencies and parents to identify students expected to be evacuated.
- Establish a keyhole evacuation to 4 miles and 170 degrees.
- Residents from 4 to 10 miles should shelter until the keyhole area has been evacuated.
- Pedestrian evacuation, or walking, should be part of the evacuation plan.

The recommendation to have a goal of 100 percent public participation is consistent with current emergency planning within EPZs. Emergency planning at the family level is also consistent with current FEMA guidance. Conducting surveys of the public that reside within 2 miles of the plant would likely provide very helpful information. This could be costly and would have to be conducted at designated time intervals for the information to remain current. The report suggests that transit dependent residents within 2 miles of the plant not be asked to wait outside for a bus. It is suggested that designated pick up points be established where shelter can be provided until the buses arrive. The logistics of establishing shelters such as those recommended were investigated in the PAR project (NRC, 2007), and it was found that: assuring 24 hour access was challenging because; people need a way to get to the facility; and the number of facilities can be quite large. However, alternatives to waiting outside along a bus route should be investigated further.

The recommendation that schools within 4 miles of the plant work more closely with emergency planners would also likely provide a benefit. Planning for 100 percent evacuation of schools is needed regardless of commitments from parents that they will evacuate their own children. The accident could occur when the parent is not at home. It is very likely that given the wide use of cell phones among children, parents will become aware of an impending school evacuation before buses are mobilized. If schools are aware of the potential number of children to evacuate, the need for second or third waves of buses may be reduced once it is confirmed that children have indeed left with a parent.

A modified keyhole evacuation area was recommended maintaining the current 2 mile, 360 degree evacuation combined with a keyhole to 4 miles instead of the current 5 miles. In addition, the keyhole shape would be expanded to 170 degrees. The 4-mile distance was based on the consequence analysis and the 170 degrees considered the potential for wind shift. The report provides a wind persistence study and shows that even with a 170 degree keyhole there is still almost a 30 percent probability that wind would shift beyond 170 degrees within 2 hours. The proposed modification of the keyhole is fairly consistent with the recent update to NUREG-0654/FEMA-REP-1, Rev. 1, Supplement 3 (NRC, 2011a) which also proposes a wind persistence study be developed for use in licensee protective action recommendations. The results of a wind persistence study would inform the licensee's recommendation showing whether additional sectors should be evacuated. Supplement 3 maintains the 5-mile distance away from the plant. The RBR report demonstrates that 4 miles could be acceptable, however, uncertainty exists throughout the accident and consequence analyses and specifying 4 miles would imply a level of precision that may be difficult to defend.

The recommendation to shelter residents from 4 to 10 miles until the keyhole has been evacuated is also consistent with the recent update to NUREG-0654/FEMA-REP-1, Rev. 1, Supplement 3. The update to Supplement 3 identifies a staged evacuate for the 2 mile area and 5 mile keyhole as the initial preferred protective action when conditions support such an evacuation.

There is an emphasis throughout the RBR report on reducing the number of evacuating vehicles. One recommendation is to encourage evacuees to walk to the 4-mile boundary and be picked up at that point by buses. Walking is often mentioned in other studies as a faster means of evacuating when ETEs show vehicle speeds of just a mile or two per hour, as is the case with the Indian Point site used in the RBR report. However, walking is not a practical means of evacuation for a nuclear power plant accident. For instance, within the site used in the RBR report, there are limited sidewalks along the evacuation routes. The terrain is hilly and these routes are not radial. Encouraging residents who may not be healthy enough to complete such a walk could result in unplanned logistical issues of trying to find and pick these residents up during an evacuation. Furthermore, if an emergency plan is developed assuming that 10 to 20 percent (an arbitrary figure) of the public is going to walk and an accident occurs during the night, adverse weather, or simply a cold winter day, the emergency plan would not have identified the resources needed to support the evacuation. Walking requires designating pickup points at the 4-mile zone and does not allow residents to bring any items with them during the evacuation. The above issues are not easily resolved and make comprehensive emergency planning difficult when trying to integrate walking into the plan.

While the staff does not agree with all the conclusions of the RBR report, the methods of calculating potential consequence are of interest and have influenced this study. These techniques begin to merge probabilistic safety analysis with EP to quantify the risk significance of individual program elements. This is analogous to identification of the most risk significant equipment for prevention of core damage and may eventually be used to stratify regulatory concern.

NRC FORM 335 (12-2010) NRCMD 3.7	U.S. NUCLEAR REGULATORY COMMISSION **BIBLIOGRAPHIC DATA SHEET** *(See instructions on the reverse)*	1. REPORT NUMBER (Assigned by NRC, Add Vol., Supp., Rev., and Addendum Numbers, if any.) NUREG/CR 7160 SAND2012-3144P

2. TITLE AND SUBTITLE	3. DATE REPORT PUBLISHED	
Emergency Preparedness Significance Quantification Process: Proof of Concept	MONTH June	YEAR 2013
	4. FIN OR GRANT NUMBER	
	R3149	

5. AUTHOR(S)	6. TYPE OF REPORT	
Randolph Sullivan, U.S. Nuclear Regulatory Commission Joseph Jones, Sandia National Laboratories	Technical	
Jeff LaChance, Sandia National Laboratories Fontini Walton, Sandia National Laboratories Scott Weber, Sandia National Laboratories	7. PERIOD COVERED (Inclusive Dates) 1/2013 until updated	

8. PERFORMING ORGANIZATION - NAME AND ADDRESS (If NRC, provide Division, Office or Region, U. S. Nuclear Regulatory Commission, and mailing address; if contractor, provide name and mailing address.)

Sandia National Laboratories, New Mexico
P.O. Box 5800
Albuquerque, NM 87185

9. SPONSORING ORGANIZATION - NAME AND ADDRESS (If NRC, type "Same as above"; if contractor, provide NRC Division, Office or Region, U. S. Nuclear Regulatory Commission, and mailing address.)
Division of Preparedness and Response
Office of Nuclear Security and Incident Response
U.S. Nuclear Regulatory Commission
Washington, DC 20555-0001

10. SUPPLEMENTARY NOTES
ML12263A317

11. ABSTRACT (200 words or less)
In an ongoing effort to increase effectiveness and efficiency through improved prioritization of regulatory activities, a decision process has been developed to aid in the determination of risk significance of Emergency Preparedness (EP) program elements. The DedUctive Quantification Index (DUQI) method was developed and used in a proof of concept application for two representative nuclear power plant sites. The results show the cumulative population dose is reduced through implementation of a formal EP program compared to conditions in which an emergency response would be implemented in an ad hoc manner. Dose was shown to be consistently lower for all analyses. The DUQI method was also applied to determine risk significance of specific EP elements. Analyses included a response where sirens are assumed not operable in the 2-5 mile area around the nuclear power plant, and for a delay in notification to offsite response organizations. Detailed consequence analysis modeling was performed using site specific information. The process used information from historical studies, such as NUREG-1150 combined with current knowledge. Data for specific sites was used in selected areas to increase the credibility of the product, but the results are not applicable to any specific site. Improvements were made to the modeling approach by simulating evacuee road loading in greater detail than previous studies. The 95th percentile cumulative population dose results were produced and used to support the study conclusions.

12. KEY WORDS/DESCRIPTORS (List words or phrases that will assist researchers in locating the report.)	13. AVAILABILITY STATEMENT
emergency preparedness	unlimited
regulatory oversight	14. SECURITY CLASSIFICATION
consequence analysis	(This Page)
significance determination process	unclassified
risk-informed	(This Report)
protective action	unclassified
consequence	15. NUMBER OF PAGES
consequence model	
	16. PRICE

NRC FORM 335 (12-2010)

UNITED STATES
NUCLEAR REGULATORY COMMISSION
WASHINGTON, DC 20555-0001

OFFICIAL BUSINESS

NUREG/CR-7160

Emergency Preparedness Significance Quantification Process: Proof of Concept

June 2013

www.ingramcontent.com/pod-product-compliance
Lightning Source LLC
Chambersburg PA
CBHW080254180526
45167CB00006B/2528